T0213753

Lecture Notes in Computer Science   8752

Commenced Publication in 1973
Founding and Former Series Editors:
Gerhard Goos, Juris Hartmanis, and Jan van Leeuwen

More information about this series at http://www.springer.com/series/7407

Pierrick Legrand · Marc-Michel Corsini
Jin-Kao Hao · Nicolas Monmarché
Evelyne Lutton · Marc Schoenauer (Eds.)

Artificial Evolution

11th International Conference,
Evolution Artificielle, EA 2013
Bordeaux, France, October 21–23, 2013
Revised Selected Papers

 Springer

Editors

Pierrick Legrand
INRIA Bordeaux sud-ouest, Equipe CQFD,
Institut de Mathématiques de Bordeaux
   (IMB), UMR CNRS 5251,
Université de Bordeaux,
3ter place de la victoire, 33076 Bordeaux,
France

Marc-Michel Corsini
Université de Bordeaux,
3ter place de la victoire, 33076 Bordeaux,
France

Jin-Kao Hao
LERIA,
Université d'Angers,
2 Bd Lavoisier, 49045 Angers Cedex 01,
France

Nicolas Monmarché
Laboratoire d'Informatique de
   l'Université de Tours,
Ecole Polytechnique de l'Université
   de Tours,
64 avenue Jean Portalis, 37200 Tours,
France

Evelyne Lutton
INRA, UMR 782 GMPA,
1 Av. Brétignière,
78850, Thirverval-Grignon,
France

Marc Schoenauer
Equipe TAO, INRIA Futurs, LRI,
Université de Paris-Sud,
Bât 490, 91405 Orsay cedex,
France

ISSN 0302-9743   ISSN 1611-3349   (electronic)
ISBN 978-3-319-11682-2   ISBN 978-3-319-11683-9   (eBook)
DOI 10.1007/978-3-319-11683-9

Library of Congress Control Number: 2014953268

LNCS Sublibrary: SL1 – Theoretical Computer Science and General Issues

Springer Cham Heidelberg New York Dordrecht London

Artificial Evolution 2013, Presented by W.B. Langdon

It was my pleasure, as an invited speaker at Artificial Evolution 2013 Bordeaux, to be invited by Pierrick Legrand to contribute this view of the conference. First, let me thank Pierrick and the whole EA 2013 conference in general for their kind invitation and wonderful hospitality. Although I saw most of the posters and presentations, it is impossible to treat them all fairly, for that, I must direct you to the individual papers, instead this will be a personal view

For me the conference started in bright sunshine, with registration in the newly completed Inria building in Bordeaux. All the sessions were held in the *Ada Lovelace* room. The first session concerned solving discrete problems with presentations by Rym Nesrine Guibadj on a delivery van route application and then Olivier Gach on detecting clusters in potentially small world neighbor connectivity graphs. During the conference several presenters said they used Pascal which provoked discussion about its merits versus the wisdom of more mainstream languages such as C and Java. The second part of the session contained presentations by Ines Sghir and Fazia Aiboud. Although they continued the discrete mathematics theme, Fazia's talk on cellular automata was motivated by the desire to model living cells within human organs.

Arnaud Liefooghe's presentation continued the discrete mathematics theme from the first session and included enumerating many benchmarks looking for features to explain their difficulty for multi-objective solvers. Sandra Astete Morales showed new proofs for the speed at which EAs will approach an optimum in continuous problems where in the neighborhood of the optimum the objective function falls as the square of the distance from it. Notice that finding the exact optimum is hard without additional assumptions as the guidance the objective function provides dies toward zero as the optimum is approached. Optima of continuous problems can often be represented by the first few terms of a Taylor expansion and thus often have this *sphere x^2*-like property. I wondered if tighter bounds might be found by using the fact that Sandra's theorems assumed Gaussian noise, whose variances can be simply summed. Charlie Vanaret's Charibde program married differential evolution with branch and bound to manage the problem in applications with many dimensions that branch and bound potentially needs to keep track of an exponentially large number of options which must still be explored.

The final session on Monday returned to evolving solutions for applications. Hoang Luong presented his work on using genetic algorithms to optimally plan extensions to the Dutch electricity distribution network (10–50 KV) on the assumption that demand for electricity will continue to grow in parts of the Netherlands. Then, Laetitia Jourdan presented work by Khedidja Seridi on using ParadisEO to try to attribute diseases, particularly human diseases, to genetic mutations (SNPs).

In the evening, Inria and the conference were opened to both the conference attendees and the members of the public for an exhibition of more than 30 evolved art pieces by 13 international artists. Many of these were visually stunning colored pictures but they also included fashionable textiles and tapestries. The exhibit commenced with Emmanuel Cayla elaborating on how he used the Evelyne Lutton's ArtiE-Fract software to interactively evolve mathematical fractal patterns and then used the tool to build on these to give vent to his artistic desires. Evelyne Lutton provided a real-time translation from French to English.

Tuesday morning we started with a nice presentation by Federic Kruger on using computer graphics hardware (GPU) within the EASEA platform to evolve predictions for electricity demand in Strasbourg. This was closely followed by Alberto Tonda's description of using evolutionary computation and user interaction to control which variables cause others (the direction of inference) in Bayesian networks. Alberto's target was Bayesian networks for use in the French food industry, particularly the creation of learning networks for cheese ripening and biscuit baking.

Prof. Jean Louis Deneubourg from the Université libre de Bruxelles' unit of social ecology (USE) gave a wonderful invited talk including the mathematical scaling laws observed in social animals, particularly cockroaches.

The poster session included new nature inspired algorithms and new approaches to new delivery van scheduling problems.

After lunch we were treated to Christian Blum's description of using ant colony optimization (ACO) in combination with beam search applied to the problem of finding the longest repetition-free subsequence in Bioinformatics strings.

No trip to Bordeaux would be complete without a visit to the wine country. Tuesday evening we were whisked through the Bordeaux region to St. Emillion. First, a very informative tour of the town itself, in which our guide pointed out the ancient fortification constructed under the orders of John "Lackland" (known in England as King John, 1166–1216) who then, but not for much longer, ruled this part of France. Since St. Emillion was claimed both by the King of France and that of England, the people of Saint Emillion tactfully refer to it as "the King's tower" without saying which king. After an unnoticed incident with the first bus, we moved seamlessly in a second to a nearby chateau (Chateau de Sarpe) and its vineyards. Arriving before sunset to see the same we received a very informative and entertaining description of its recent history and viticulture before retiring to its ground level cellars to see both the current and aging vintages and hear more of their production. Many estimates were made of the value of the fluid concealed within the oaken barrels before ample opportunity to sample the same stiffened with numerous fine French cheeses and preserved meats.

Wednesday morning, wishing to refresh my presentation slides, I was late and so have little report on the second theory session. Nonetheless after it, in the genetic programming session, Nuno Lourenco gave a nice discursive talk on using an evolutionary algorithm, grammatical evolution, to evolve the selection step in another evolutionary algorithm. He applied the resulting hyper-heuristic to knapsack problems and also described extensions he made to the traditional Backus-Naur Form (BNF) grammar used in GE.

After my talk, we proceeded on foot happily as usual to the nearby Brasserie "Le 7eme Art" for lunch. During dessert the heavens opened and we were treated to a fine display of lightning with thunder accompaniment. As might be guessed no one was keen to run back to the Inria building. Instead my suggestion that we put the show on in the restaurant was met with approval by all[1] except, surprisingly, by Alina Mereuta, who was to be the first speaker. Due to heroic efforts by Pierrick, Evelyne, Laetitia Grimaldi, and Ingrid Rochel and the Inria EA 2013 conference souvenir umbrellas we did indeed return to Inria and Alina extracted her revenge. Nonetheless, her talk on using CMA-ES in real time to optimally rebalance the colours used on web pages to best alleviate the effects of color blindness was well received and provoked a lively discussion. Which included many suggestions such as: caching answers for the same or similar pages; treating CMA as an anytime algorithm and updating the display each time it finds a better answer, rather than waiting for it to find the optimal answer; and allowing the user to instantly undo changes to prevent other people catching sight of his screen and then re-apply changes after the boss has left the room. She was followed by Juan Carlos Rivera who described work on emergency management, especially scheduling disaster relief supplies.

The conference was concluded by Alejandro Lopez Rincon describing inferring problems with the human heart from skin voltage measurements taken by an array of electrical contacts on the chest and back and Piero Consoli and Mario Pavone's paper on partitioning networks.

November 2013   W.B. Langdon

---

[1] Who have given their presentations.

Preface

This LNCS volume includes the best papers presented at the 11th Biennial International Conference on Artificial Evolution, EA^2 2013, held in Bordeaux (France). Previous EA editions took place in Angers (2011), Strasbourg (2009), Tours (2007), Lille (2005), Marseille (2003), Le Creusot (2001), Dunkerque (1999), Nimes (1997), Brest (1995), and Toulouse (1994).

Authors had been invited to present original work relevant to Artificial Evolution, including, but not limited to: Evolutionary Computation, Evolutionary Optimization, Co-evolution, Artificial Life, Population Dynamics, Theory, Algorithmics and Modeling, Implementations, Application of Evolutionary Paradigms to the Real World (industry, biosciences, ...), other Biologically Inspired Paradigms (Swarm, Artificial Ants, Artificial Immune Systems, Cultural Algorithms...), Memetic Algorithms, Multi-Objective Optimization, Constraint Handling, Parallel Algorithms, Dynamic Optimization, Machine Learning, and hybridization with other soft computing techniques.

Each submitted paper was reviewed by four members of the International Program Committee. Among the 39 submissions received, 20 papers were selected for oral presentation and 2 other papers for poster presentation. For the previous editions, a selection of the best papers which were presented at the conference and further revised were published (see LNCS volumes 1063, 1363, 1829, 2310, 2936, 3871, 4926, 5975, and 7401). Exceptionally, for this edition, the high quality of the 20 papers selected for the oral presentation led us to include a revised version of all these paper in this volume of Springer's LNCS series.

We would like to express our sincere gratitude to our invited speakers: Jean-Louis Deneubourg and William Langdon.

The success of the conference resulted from the input of many people to whom I would like to express my appreciation: The members of Program Committee and the secondary reviewers for their careful reviews that ensure the quality of the selected papers and of the conference. The members of the Organizing Committee for their efficient work and dedication assisted by Laetitia Grimaldi, Nicolas Jahier, Cathy Metivier, and Ingrid Rochel. The members of the Steering Committee for their valuable assistance. Aurélien Dumez for his support on the administration of the website. Mélanie Toto for the design and the visual identity of the conference. Marc Schoenauer for his support with the MyReview system. Laetitia Jourdan, Lola Kovacic, and Marion Bachelet for the publicity. Marc-Michel Corsini for the edition of the proceedings. Sebastien Verel for the registrations. Evelyne Lutton and Nicolas Monmarché for the organization of the Side Event: "Art and Artificial Evolution," and the artists. Laurent Vezard for his support.

---

^{2} As for previous editions of the conference, the EA acronym is based on the original French name "Évolution Artificielle."

I take this opportunity to thank the different partners whose financial and material support contributed to the organization of the conference: Université Bordeaux 1, Université Bordeaux Segalen, Région Aquitaine, La CUB, CNRS, IMB, Inria, UFR Sciences et Modélisation.

Last but not least, I thank all the authors who submitted their research papers to the conference, and the authors of accepted papers who attended the conference to present their work. Thank you all.

October 2013   Pierrick Legrand

Evolution Artificielle 2013 – EA 2013

October 21–23, 2013
Bordeaux, France
11th International Conference on Artificial Evolution

Program Committee

Sebastien Aupetit	Université de Tours, France
Nicolas Bredeche	Université Pierre et Marie Curie, France
Stefano Cagnoni	Universita di Parma, Italy
Pierre Collet	Université de Strasbourg, France
Marc-Michel Corsini	Université de Bordeaux, France
Benjamin Doerr	Max-Planck-Institut für Informatik, Germany
Nicolas Durand	IRIT, France
Marc Ebner	Ernst-Moritz-Arndt-Universität Greifswald, Germany
Cyril Fonlupt	ULCO, France
Christian Gagné	Université Laval, Canada
Edgar Galvan	Trinity College Dublin, Ireland
Mario Giacobini	University of Turin, Italy
Jens Gottlieb	SAP AG, Germany
Eric Grivel	IMS, France
Jin-Kao Hao	Université d'Angers, France
Colin Johnson	University of Kent, UK
Laetitia Jourdan	LIFL, France
Pierrick Legrand	Université de Bordeaux, France
Jean Louchet	Ghent University, Belgium
Evelyne Lutton	INRA-AgroParisTech, France
Virginie Marion-Poty	LISIC, ULCO, France
Nicolas Monmarché	University of Tours, France
Jean-Marc Montanier	CRAB Team, NTNU, Norway
Gabriela Ochoa	University of Stirling, UK
Denis Robilliard	LISIC, ULCO, France
Eduardo Rodriguez-Tello	CINVESTAV, Tamaulipas, Mexico
Frédéric Saubion	Université d'Angers, France
Marc Schoenauer	Inria Saclay, France
Patrick Siarry	Université Paris-Est Créteil, France
Christine Solnon	Université Lyon 1, France
Thomas Stuetzle	Université libre de Bruxelles, Belgique
El-Ghazali Talbi	University of Lille, France
Olivier Teytaud	TAO, Inria, France
Fabien Teytaud	LISIC, France

Alberto Tonda   UMR 782 GMPA INRA, France
José Torres-Jiménez   CINVESTAV, Tamaulipas, Mexico
Leonardo Trujillo   ITT, Mexico
Shigeyoshi Tsutsui   Hannan University, Japan
Paulo Urbano   University of Lisbon, Portugal
Gilles Venturini   University of Tours, France
Sébastien Verel   ULCO, France
Laurent Vezard   CQFD, Inria, France

Steering Committee

Pierre Collet   Université Louis Pasteur de Strasbourg, France
Jin-Kao Hao   Université d'Angers, France
Evelyne Lutton   INRA, France
Nicolas Monmarché   Université François Rabelais de Tours, France
Marc Schoenauer   Inria, France

Organizing Committee

Marc-Michel Corsini (LNCS Publication, Proceedings)
Aurelien Dumez (Admin Web)
Laetitia Grimaldi (Local Organization)
Nicolas Jahier (Local Organization)
Laetitia Jourdan (Publicity)
Pierrick Legrand (General Chair, LNCS Publication, Proceedings)
Ingrid Rochel (Local Organization)
Marc Schoenauer (Submissions, MyReview)
Mélanie Toto (Webdesigner)
Sébastien Vérel (Treasurer)
Laurent Vezard (Local Organization)

Invited Speakers

Jean Louis Deneubourg, Université libre de Bruxelles

Exploring animal collective behaviors

Jean-Louis Deneubourg is a FNRS researcher and the director of the Social Ecology Department (Faculté des Sciences ULB). He is a member of the Royal Academy of Belgium. His research activities are oriented toward collectives behaviors in animal societies and artificial systems. He has written 300 publications on these topics.

Photo by Nicolas Monmarché

William B. Langdon, University College London, Department of Computer Science

Genetic Improvement Programming

Bill Langdon gained his PhD at University College, London after a career in industrial control software and IT consulting. He has held positions in universities and research institutes both in England and overseas. Recently, he has applied genetic programming to optimizing software. He has written three books on genetic programming, including *A Field Guide to Genetic Programming* and maintains the Genetic Programming Bibliography.

Photo by Nicolas Monmarché

Side Event

Evolutionary Artworks Exhibition – *October 21, 2013, 17h45*

This exhibition, open to the public, displayed a series of artworks created by artists and researchers using approaches based on artificial evolution.

The "biodiversity" of evo-artists was illustrated by works of a dozen different authors. Some of them were on place for discussions, and an interactive design show was presented by the artist Emmanuel Cayla. An electronic artworks catalog is available on the conference website.

http://ea2013.inria.fr//EA2013-catalogue_side-event_A3.pdf

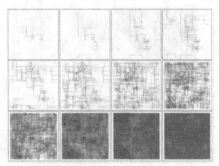

Nicolas Monmarché – *Ants painting.*
Artificial Ants

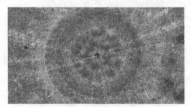

Alejandro Lopez Rincon

Brian J. Ross – Maryam Baniasadi – Cherries
Genetic Programming

Günter Bachelier
Evolutionary Art History Appropriation

Organization and Contact:
– Evelyne Lutton (evelyne.lutton@grignon.inra.fr)
– Nicolas Monmarché (nicolas.monmarche@univ-tours.fr)

Contents

Theory

What Makes an Instance Difficult for Black-Box 0–1 Evolutionary Multiobjective Optimizers?

Arnaud Liefooghe[1,2](✉), Sébastien Verel[3], Hernán Aguirre[4],
and Kiyoshi Tanaka[4]

[1] LIFL (UMR CNRS 8022), Université Lille 1,
Villeneuve d'Ascq, France
[2] Dolphin, Inria Lille-Nord Europe, Villeneuve d'Ascq, France
arnaud.liefooghe@univ-lille1.fr
[3] LISIC, Université du Littoral Côte d'Opale, Calais, France
verel@lisic.univ-littoral.fr
[4] Faculty of Engineering, Shinshu University, Nagano, Japan
{ahernan,ktanaka}@shinshu-u.ac.jp

Abstract. This paper investigates the correlation between the characteristics extracted from the problem instance and the performance of a simple evolutionary multiobjective optimization algorithm. First, a number of features are identified and measured on a large set of enumerable multiobjective NK-landscapes with objective correlation. A correlation analysis is conducted between those attributes, including low-level features extracted from the problem input data as well as high-level features extracted from the Pareto set, the Pareto graph and the fitness landscape. Second, we experimentally analyze the (estimated) running time of the global SEMO algorithm to identify a $(1 + \varepsilon)$-approximation of the Pareto set. By putting this performance measure in relation with problem instance features, we are able to explain the difficulties encountered by the algorithm with respect to the main instance characteristics.

1 Introduction

In single-objective black-box combinatorial optimization, fitness landscape analysis aims at apprehending the relation between the geometry of a problem instance and the dynamics of randomized search algorithms. Understanding the main problem-related features allows to explain the behavior and the performance of such algorithms, the ultimate goal being to predict this performance and adapt the algorithm setting to the instance being solved. Recently, the performance of single-objective randomized search algorithms has been correlated to fitness landscape features [2]. In this paper, we propose a general methodology to analyze the correlation between problem features and algorithm performance in black-box 0–1 evolutionary multiobjective optimization. To the best of our knowledge, this is the first time that such an analysis is conducted in multiobjective optimization.

© Springer International Publishing Switzerland 2014
P. Legrand et al. (Eds.): EA 2013, LNCS 8752, pp. 3–15, 2014.
DOI: 10.1007/978-3-319-11683-9_1

We first identify a number of existing and original multiobjective problem features. They include low-level features extracted from the problem input data like variable correlation, objective correlation, and objective space dimension, as well as high-level features from the Pareto set, the Pareto graph and the ruggedness and multimodality of the fitness landscape. Some of them are here proposed for the first time. They consist of a simple autocorrelation function, based on a local hypervolume measure, and allowing to estimate the ruggedness of the fitness landscape. We report all these measures on a large number of enumerable multiobjective NK-landscapes with objective correlation (ρMNK-landscapes), together with a correlation analysis between them.

Next, we conduct an experimental analysis on the correlation between instance features and algorithm performance. To do so, we investigate the estimated running time of a simple evolutionary multiobjective optimization algorithm, namely global SEMO [7], to identify a $(1 + \varepsilon)$-approximation of the Pareto set. In particular, the original hypervolume-based autocorrelation functions appear to be the features with the highest correlation with the algorithm performance. Overall, the running time of the algorithm is impacted by each of the identified multiobjective problem feature. Our analysis shows their relative importance on the algorithm efficiency. Moreover, taking the features all together allows to better explain the dynamics of randomized search algorithms.

The paper is organized as follows. Section 2 details the background information related to fitness landscape analysis, multiobjective optimization and ρMNK-landscapes. In Sect. 3, low-level and high-level instance features are identified, and quantitative results, together with a correlation analysis, are reported for ρMNK-landscapes. Section 4 presents the experimental setup of global SEMO and discusses the correlation between the problem features and the estimated running time of global SEMO. Section 5 concludes the paper and discusses further research.

2   Preliminaries

2.1   Fitness Landscape Analysis

In single-objective optimization, fitness landscape analysis allows to study the topology of a combinatorial optimization problem [13], by gathering important information such as ruggedness or multimodality. A fitness landscape is defined by a triplet $(X, \mathcal{N}, \phi)$, where X is a set of admissible solutions (the search space), $\mathcal{N} : X \to 2^X$ is a neighborhood relation, and $\phi : X \to \mathbb{R}$ is a (scalar) fitness function, here assumed to be maximized. A *walk* over the fitness landscape is an ordered sequence $\langle x_0, x_1, \ldots, x_\ell \rangle$ of solutions from the search space such that $x_0 \in X$, and $x_t \in \mathcal{N}(x_{t-1})$ for all $t \in \{1, \ldots, \ell\}$.

An *adaptive walk* is a walk such that for all $t \in \{1, \ldots, \ell\}$, $\phi(x_t) > \phi(x_{t-1})$, as performed by a conventional hill-climbing algorithm. The number of iterations, or steps, of the hill-climbing algorithm is the length of the adaptive walk. This length is a good estimator of the average diameter of the local optima basins of attraction, characterizing a problem instance multimodality. The larger the length, the larger the basin diameter. This allows to estimate the number of local optima when the whole search space cannot be enumerated exhaustively.

Let $\langle x_0, x_1, \ldots \rangle$ be an infinite *random walk* over the search space. The autocorrelation function and the correlation length of such a random walk allow to measure the ruggedness of a fitness landscape [13]. The random walk autocorrelation function $r : \mathbb{N} \to \mathbb{R}$ of a (scalar) fitness function ϕ is defined as follows.

$$r(k) = \frac{\mathbb{E}[\phi(x_t) \cdot \phi(x_{t+k})] - \mathbb{E}[\phi(x_t)] \cdot \mathbb{E}[\phi(x_{t+k})]}{\text{Var}(\phi(x_t))} \tag{1}$$

where $\mathbb{E}[\phi(x_t)]$ and $\text{Var}(\phi(x_t))$ are the expected value and the variance of $\phi(x_t)$, respectively. The autocorrelation coefficients $r(k)$ can be estimated within a finite random walk $\langle x_0, x_1, \ldots, x_\ell \rangle$ of length ℓ.

$$\hat{r}(k) = \frac{\sum_{t=1}^{\ell-k} (\phi(x_t) - \bar{\phi}) \cdot (\phi(x_{t+k}) - \bar{\phi})}{\sum_{t=1}^{\ell} (\phi(x_t) - \bar{\phi})^2} \tag{2}$$

where $\bar{\phi} = \frac{1}{\ell} \sum_{t=1}^{\ell} \phi(x_t)$, and $\ell \gg 0$. The estimation error diminishes with the walk length ℓ. The correlation length τ measures how the autocorrelation function decreases. This characterizes the ruggedness of the landscape: the larger the correlation length, the smoother the landscape. Following [13], we define the correlation length by $\tau = -\frac{1}{\ln(r(1))}$, making the assumption that the autocorrelation function decreases exponentially.

2.2 Multiobjective Optimization

A *multiobjective optimization problem* can be defined by an objective vector function $f = (f_1, \ldots, f_M)$ with $M \geqslant 2$ objective functions, and a set X of feasible solutions in the *decision space*. In the combinatorial case, X is a discrete set. Let $Z = f(X) \subseteq \mathbb{R}^M$ be the set of feasible outcome vectors in the *objective space*. To each solution $x \in X$ is assigned an objective vector $z \in Z$ on the basis of the vector function $f : X \to Z$ with $z = f(x)$. The conventional Pareto dominance relation is defined as follows. In a maximization context, an objective vector $z \in Z$ is dominated by an objective vector $z' \in Z$, denoted by $z \prec z'$, if and only if $\forall m \in \{1, \ldots, M\}$, $z_m \leqslant z'_m$ and $\exists m \in \{1, \ldots, M\}$ such that $z_m < z'_m$. By extension, a solution $x \in X$ is dominated by a solution $x' \in X$, denoted by $x \prec x'$, if and only if $f(x) \prec f(x')$. A solution $x^\star \in X$ is said to be *Pareto optimal* (or efficient, non-dominated), if and only if there does not exist any other solution $x \in X$ such that $x^\star \prec x$. The set of all Pareto optimal solutions is called the *Pareto set* $X^\star \subseteq X$. Its mapping in the objective space is called the *Pareto front* $Z^\star \subseteq Z$. One of the most challenging task in multiobjective optimization is to identify a minimal complete Pareto set [3], *i.e.* a Pareto set of minimal size, that is one Pareto optimal solution for each point from the Pareto front.

However, in the combinatorial case, generating a complete Pareto set is often infeasible for two main reasons [3]: (*i*) the number of Pareto optimal solutions is typically exponential in the size of the problem instance, and (*ii*) deciding if a feasible solution belongs to the Pareto set may be NP-complete. Therefore, the overall goal is often to identify a good *Pareto set approximation*. To this end, heuristics in general, and evolutionary algorithms in particular, have received a growing interest since the late eighties.

2.3 ρMNK-Landscapes

The family of ρMNK-landscapes constitutes a problem-independent model used for constructing multiobjective multimodal landscapes with objective correlation [12]. It extends single-objective NK-landscapes [6] and multiobjective NK-landscapes with independent objective functions [1]. Feasible solutions are binary strings of size N, *i.e.* the decision space is $X = \{0, 1\}^N$. The parameter N refers to the problem size (the bit-string length), and the parameter K to the number of variables that influence a particular position from the bit-string (the epistatic interactions). The objective vector function $f = (f_1, \ldots, f_m, \ldots, f_M)$ is defined as $f : \{0, 1\}^N \to [0, 1)^M$. Each objective function f_m is to be maximized and can be formalized as follows.

$$f_m(x) = \frac{1}{N} \sum_{i=1}^{N} c_i^m(x_i, x_{i_1}, \ldots, x_{i_K}), m \in \{1, \ldots, M\} \tag{3}$$

where $c_i^m : \{0, 1\}^{K+1} \to [0, 1)$ defines the multidimensional component function associated with each variable x_i, $i \in \{1, \ldots, N\}$, and where $K < N$. By increasing the number of variable interactions K from 0 to $(N - 1)$, ρMNK-landscapes can be gradually tuned from smooth to rugged. In this work, we set the position of these epistatic interactions uniformly at random. The same epistatic degree $K_m = K$ and the same epistatic interactions are used for all objectives $m \in \{1, \ldots, M\}$. Component values are uniformly distributed in the range $[0, 1)$, and follow a multivariate uniform distribution of dimension M, defined by a correlation coefficient $\rho > \frac{-1}{M-1}$, *i.e.* the same correlation ρ is defined between all pairs of objective functions. As a consequence, it is very unlikely that the same objective vector is assigned to two different solutions. The positive (respectively negative) data correlation allows to decrease (respectively increases) the degree of conflict between the objective function values very precisely [12]. An instance generator and the problem instances under study in this paper can be found at the following URL: http://mocobench.sf.net/.

In the following, we investigate ρMNK-landscapes with an epistatic degree $K \in \{2, 4, 6, 8, 10\}$, an objective space dimension $M \in \{2, 3, 5\}$, and an objective correlation $\rho \in \{-0.9, -0.7, -0.4, -0.2, 0.0, 0.2, 0.4, 0.7, 0.9\}$ such that $\rho > \frac{-1}{M-1}$. The problem size is set to $N = 18$ in order to enumerate the search space exhaustively. The search space size is then $|X| = 2^{18}$. 30 different landscapes, independently generated at random, are considered for each parameter combination: ρ, M, and K. This leads to a total of 3300 problem instances.

3 Problem Features and Correlation Analysis

In this section, we identify a number of general-purpose features, either directly extracted from the problem instance itself (low-level features), or computed from the enumerated Pareto set and from the fitness landscape (high-level features). Then, a correlation analysis is conducted on those features in order to highlight the main similarities in characterizing the difficulties of a problem instance.

3.1 Low-Level Features from Problem Input Data

First, we consider some features related to the definition of ρMNK-landscapes.

Number of epistatic interactions (K): This gives the number of variable correlations in the construction of ρMNK-landscapes. As will be detailed later, despite the K-value can generally not be retrieved directly from an unknown instance, it can be precisely estimated within some high-level fitness landscape metrics described below.

Number of objective functions (M): This parameter represents the dimension of the objective space in the construction of ρMNK-landscapes.

Objective correlation (ρ): This parameter allows to tune the correlation between the objective function values in ρMNK-landscapes. In our analysis, the objective correlation is the same between all pairs of objectives.

3.2 High-Level Features from the Pareto Set

The high-level fitness landscape metrics considered in our analysis are described below. We start with some general features related to the Pareto set.

Number of Pareto optimal solutions (npo): The number of Pareto optimal solutions enumerated in the instance under consideration simply corresponds to the cardinality of the (exact) Pareto set, *i.e.* $npo = |X^\star|$. The approximation set manipulated by any EMO algorithm is directly related to the cardinality of the Pareto optimal set. For ρMNK-landscapes, the number of Pareto optimal solutions typically grows exponentially with the problem size, the number of objectives and with the degree of conflict between the objectives [12].

Hypervolume (hv): The hypervolume value of a the Pareto set $X^\star$ gives the portion of the objective space that is dominated by $X^\star$ [14]. We take the origin as a reference point $z^\star = (0.0, \ldots, 0.0)$.

Average distance between Pareto optimal solutions (avgd): This metric corresponds to the average distance, in terms of Hamming distance, between any pair of Pareto optimal solutions.

Maximum distance between Pareto optimal solutions (maxd): This metric is the maximum distance between two Pareto optimal solutions in terms of Hamming distance.

3.3 High-Level Features from the Pareto Graph

In the following, we describe some high-level features related to the *connectedness* of the Pareto set [4]. If all Pareto optimal solutions are connected with respect to a given neighborhood structure, the Pareto set is said to be *connected*, and local search algorithms would be able to identify many non-dominated solutions by starting with at least one Pareto optimal solution; see *e.g.* [9,10]. We follow the definition of *k-Pareto graph* from [9]. The k-Pareto graph is defined as a graph $PG_k = (V, E)$, where the set of vertices V contains all Pareto optimal solutions, and there is an edge $e_{ij} \in E$ between two nodes i and j if and only if the shortest distance between solutions x_i and $x_j \in X$ is below a bound k, *i.e.* $d(x_i, x_j) \leqslant k$. The distance $d(x_i, x_j)$ is taken as the Hamming distance for ρMNK-landscapes. This corresponds to the *bit-flip* neighborhood operator. Some connectedness-related high-level features under investigation are given below.

Number of connected components (nconnec): This metric gives the number of connected components in the 1-Pareto graph, *i.e.* in PG_k with $k = 1$.

Size of the largest connected component (lconnec): This corresponds to the size of the largest connected component in the 1-Pareto graph PG_1.

Minimum distance to be connected (kconnec): This measure corresponds to the smallest distance k such that the k-Pareto graph is connected, *i.e.* for all pairs of vertices $x_i, x_j \in V$ in PG_k, there exists and edge $e_{ij} \in E$.

3.4 High-Level Features from the Fitness Landscape

At last, we give some high-level metrics related to the number of local optima, the length of adaptive walks, and the autocorrelation functions.

Number of Pareto local optima (nplo): A solution $x \in X$ is a *Pareto local optimum* with respect to a neighborhood structure $\mathcal{N}$ if there does not exist any neighboring solution $x' \in \mathcal{N}(x)$ such that $x \prec x'$; see *e.g.* [11]. For ρMNK-landscapes, the neighborhood structure is taken as the *1-bit-flip*, which is directly related to a Hamming distance 1. This metric reports the number of Pareto local optima enumerated on the ρMNK-landscape under consideration.

Length of a Pareto-based adaptive walk (ladapt): We here compute the length of adaptive walks by means of a very basic single solution-based *Pareto-based Hill-Climbing* (PHC) algorithm. The PHC algorithm is initialized with a random solution. At each iteration, the current solution is replaced by a random dominating neighboring solution. As a consequence, PHC stops on a Pareto local optimum. The number of iterations, or steps, of the PHC algorithm is the length of the Pareto-based adaptive walk. As in the single-objective case, the number of Pareto local optima is expected to increase exponentially when the adaptive length decreases for ρMNK-landscapes [12].

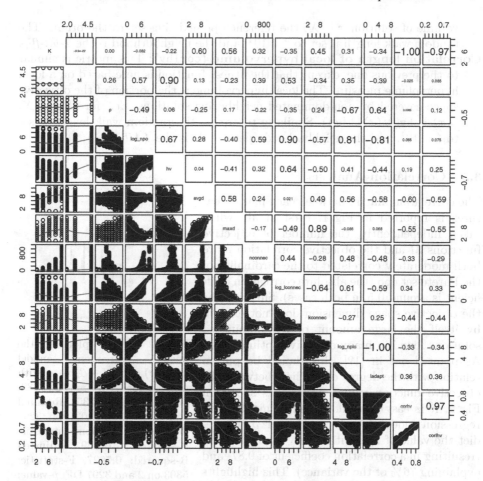

Fig. 1. Correlation matrix between all pairs of features. The feature names are reported on the diagonal. For each pair of features, scatter plots and smoothing splines are displayed below the diagonal, and the corresponding correlation coefficients are reported above the diagonal. The smoothing spline is a smoothing method that fits a smooth curve to a set of noisy observations using a spline function. The correlation coefficient is based on a Pearson product-moment correlation coefficient measuring the linear correlation (dependence) between both features. Correlation coefficient values lie between −1 (total negative correlation) and +1 (total positive correlation), while 0 means *no* correlation.

Correlation length of solution hypervolume (corhv): The ruggedness is here measured in terms of the autocorrelation of the hypervolume along a random walk. As explained in Sect. 2.1, the correlation length τ measures how the autocorrelation function, estimated with a random walk, decreases. The autocorrelation coefficients are here computed with the following scalar fitness function $\phi : X \to \mathbb{R}$: $\phi(x) = \text{hv}(\{x\})$, where $\text{hv}(\{x\})$ is the hypervolume of solution $x \in X$, the reference point being set to the origin. The random walk length is set to $\ell = 10^4$, and the neighborhood is the *1-bit-flip*.

Correlation length of local hypervolume (corlhv): This metric is similar to the previous one, except that the fitness function is here based on a local hypervolume measure. The local hypervolume is the portion of the objective space covered by non-dominated neighboring solutions, *i.e.* for all $x \in X$, $\phi(x) = \text{hv}(\mathcal{N}(x) \cup \{x\})$. Similarly to corhv, the random walk length is set to $\ell = 10^4$, and the neighborhood operator $\mathcal{N}$ is the *1-bit-flip*.

3.5 Correlation Analysis

The correlation matrix between each pair of features is reported in Fig. 1. First of all, when taken independently, the number of objective functions M and the objective correlation ρ are both moderately correlated to the cardinality of the Pareto set npo (the absolute correlation coefficient is around 0.5 in both cases). Surprisingly, the objective space dimension does not explain by itself the large amount of non-dominated solutions found in many-objective optimization. As pointed out in [12], this should be put in relation with the degree of conflicts between the objective function values. Indeed, as shown in Fig. 2, it is easy to build a simple multi-linear regression model based on M and ρ to predict the value of npo with a very high precision (resulting in a correlation coefficient of 0.87, and explaining 76 % of the variance). This highlights that the impact of many-objective fitness landscapes on the search process cannot be analyzed

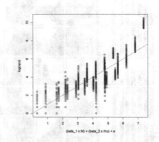

Fig. 2. Scatter plot of the linear regression model $\log(\text{npo}) = \beta_1 M + \beta_2 \rho + e$, with $\beta_1 = 1.30567$, $\beta_2 = -2.87688$, and $e = 0.27735$. Residual standard error: 1.037 on 3297 degrees of freedom, multiple R-squared: 0.7629, adjusted R-squared: 0.7627, F-statistic: 5303 on 2 and 3297 DF, p-value: $< 2.2e - 16$.

properly without taking the objective correlation into account.

Interestingly, other important remarks can be extracted from the figure. With respect to the Pareto set, the hypervolume value increases with the objective space dimension. Moreover, and unsurprisingly, the Pareto set size and the size of the largest connected component from the Pareto graph are highly correlated. So are the maximum distance between Pareto optimal solutions and the minimum distance for the Pareto set to be connected. As also reported in [12], there is a high correlation between the number of Pareto optimal solutions npo and of Pareto local optima nplo. More importantly, the number of Pareto local optima nplo can be precisely estimated with the length of a Pareto-based adaptive walk ladapt (the absolute correlation coefficient between $\log(\text{nplo})$ and ladapt is 1). As a consequence, this allows to estimate the size of the Pareto set as well. At last, the number of epistatic interactions (decision variable correlations) K can be estimated with hypervolume-based autocorrelation functions along a random walk corhv and corlhv. Since there is not much difference

---

Algorithm 1. Pseudo-code of G-SEMO

Input: $x^0 \in X$
Output: Archive A
1: $A \leftarrow x^0$
2: **loop**
3:    select x from A at random
4:    create x' by flipping each bit of x with a probability $1/N$
5:    $A \leftarrow$ non-dominated solutions from $A \cup \{x'\}$
6: **end loop**

---

between the correlations coefficients of both functions, the first one `corhv` should preferably be considered due to its simplicity. Notice that a similar analysis involving instances with the same number of objectives resulted in comparable results.

4   Problem Features *vs.* Algorithm Performance

4.1   Experimental Setup

Global SEMO.   Global SEMO (G-SEMO for short) [7] is a simple elitist steady-state EMO algorithm for black-box 0–1 optimization problems dealing with an arbitrary objective vector function defined as $f : \{0,1\}^N \rightarrow Z$ such that $Z \subset \mathbb{R}^M$, like ρMNK-landscapes. A pseudo-code is given in Algorithm 1. It maintains an unbounded archive A of non-dominated solutions found so far. The archive is initialized with one random solution from the search space. At each iteration, one solution is chosen at random from the archive. Each bit of this solution is independently flipped with a rate $r = 1/N$, and the obtained solution is checked for insertion in the archive. Within such an independent bit-flip mutation, any solution from the search space can potentially be reached by applying the mutation operator to any arbitrary solution. In its general form, the G-SEMO algorithm does not have any explicit stopping rule [7]. In this paper, we are interested in its running time, in terms of a number of function evaluations, until an $(1 + \varepsilon)$-approximation of the Pareto set has been identified and is contained in the internal memory A of the algorithm, subject to a maximum number of function evaluations.

Performance Evaluation.   For any constant value $\varepsilon \geqslant 0$, the (multiplicative) ε-dominance relation $\preceq_\varepsilon$ can be defined as follows. For all $z, z' \in Z$, $z \preceq_\varepsilon z'$ if and only if $z_m \cdot (1 + \varepsilon) \leqslant z'_m$, $\forall m \in \{1, \dots, M\}$. Similarly, for all $x, x' \in X$, $x \preceq_\varepsilon x'$ if and only if $f(x) \preceq_\varepsilon f(x')$. Let $\varepsilon \geqslant 0$. A set $X^\varepsilon \subseteq X$ is an $(1 + \varepsilon)$-approximation of the Pareto set if and only if, for any solution $x \in X$, there is one solution $x' \in X^\varepsilon$ such that $x \preceq_\varepsilon x'$. This is equivalent of finding a Pareto set approximation whose multiplicative epsilon quality indicator value with respect to the exact Pareto set is $(1 + \varepsilon)$, see *e.g.* [14]. Interestingly, under some general assumptions, there always exists an $(1 + \varepsilon)$-approximation, for any given $\varepsilon \geqslant 0$, whose cardinality is both polynomial in the problem size and in $1/\varepsilon$ [8].

Following a conventional methodology from single-objective continuous black-box optimization benchmarking [5], the expected number of function evaluations to identify an $(1 + \varepsilon)$-approximation is chosen as a performance measure. However, as any EMO algorithm, G-SEMO can either succeed or fail to reach an accuracy of ε in a single simulation run. In case of a success, the runtime is the number of function evaluations until an $(1+\varepsilon)$-approximation was found. In case of a failure, we simply restart the algorithm at random. We then obtain a *"simulated runtime"* [5] from a set of given trials of G-SEMO on a given instance. Such a performance measure allows to take into account both the success rate $p_s \in (0, 1]$ and the convergence speed of the G-SEMO algorithm. Indeed, after $(n - 1)$ failures, each one requiring T_f evaluations, and the final successful run with T_s evaluations, the total runtime is $T = \sum_{i=1}^{n-1} T_f + T_s$. By taking the expectation value and by considering that the probability of success after $(n-1)$ failures follows a Bernoulli distribution of parameter p_s, we have:

$$\mathbb{E}[T] = \left(\frac{1 - p_s}{p_s} \right) \mathbb{E}[T_f] + \mathbb{E}[T_s] \tag{4}$$

In our case, the success rate p_s is estimated with the ratio of successful runs over the total number of executions $(\hat{p}_s)$, the expected runtime for unsuccessful runs $\mathbb{E}[T_f]$ is set to a constant function evaluation limit T_{max}, and the expected runtime for successful runs $\mathbb{E}[T_s]$ is estimated with the average number of function evaluations performed by successful runs.

$$\texttt{ert} = \left(\frac{1 - \hat{p}_s}{\hat{p}_s} \right) T_{max} + \frac{1}{N_s} \sum_{i=1}^{N_s} T_i \tag{5}$$

where N_s is the number of successful runs, and T_i is the number of evaluations required for successful run i. For more details, we refer to [5].

Parameter Setting. In our analysis, we set $\varepsilon = 0.1$. The time limit is set to $T_{max} = 2^N/10 < 26215$ function evaluations without identifying an $(1 + \varepsilon)$-approximation. The G-SEMO algorithm is executed 100 times *per* instance. For a given instance, the success rate and the expected number of evaluations for successful runs are estimated from those 100 executions. However, let us note that G-SEMO was not able to identify a $(1 + \varepsilon)$-approximation set for any of the runs on one instance with $M = 3$, $\rho = 0.2$ and $K = 10$, one instance with $M = 3$, $\rho = 0.4$ and $K = 10$, ten instances with $M = 5$, $\rho = 0.2$ and $K = 10$, six instances with $M = 5$, $\rho = 0.4$ and $K = 10$, as well as two instances with $M = 5$, $\rho = 0.7$ and $K = 10$. Moreover, G-SEMO was not able to solve the following instances due to an overload CPU resources available: $M = 5$ and $\rho \in \{-0.2, 0.0\}$. Overall, this represents a total amount of 2980 instances times 100 executions, that is 298000 simulation runs.

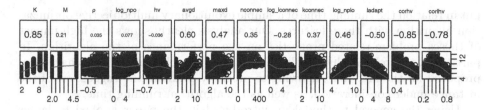

Fig. 3. Correlation between log(ert) and each feature. The feature names are reported on the first line, correlation coefficients are reported on the second line, and scatter plots as well as smoothing splines are displayed on the third line.

4.2 Computational Results

The correlation between each feature and the running time of G-SEMO is reported in Fig. 3. First, with respect to low-level features, there exists a high correlation between log(ert) and K, which is the highest absolute correlation observed on our data. However, surprisingly, the correlation of the performance measure with M and ρ is not significant. Second, with respect to high-level features from the Pareto set, the size of the Pareto set and its hypervolume does not explain the variance of log(ert). Nevertheless, the larger the distance between Pareto optimal solutions in the decision space, the larger the running time of G-SEMO. Similarly, when the Pareto graph is close to a fully connected graph, G-SEMO is likely to take less time to identify a $(1+\varepsilon)$-approximation (the absolute correlation value is around 0.3). As a consequence, the number of Pareto optimal solutions has a smaller impact on the performance of G-SEMO than the structure existing between those solutions in the decision space.

With respect to high-level fitness landscape features, the number of Pareto local optima nplo and its estimator ladapt both present a significant correlation with the estimated running time of G-SEMO. Indeed, the more Pareto local optima, the longer the running time (the absolute correlation value is close to 0.5). At last, the hypervolume-based autocorrelation functions highly explain the variance of the G-SEMO performance. For both corhv and corlhv, the absolute correlation value is around 0.8. Overall, this correlation analysis gives a "big picture" of a well-suited multiobjective fitness landscape for G-SEMO. This corroborates the impact of the problem instance properties identified in the previous section on the performance of multiobjective evolutionary algorithms.

5 Discussion

In this paper, we attempted to give a first step towards a better understanding of the evolutionary multiobjective optimization algorithm performance according to the main characteristics of the problem instance. We first presented a number of general problem features, together with a correlation analysis between those features on a large set of enumerable multiobjective NK-landscapes. Then, we put in relation the running time of a simple evolutionary multiobjective optimization algorithm with those features. Our analysis clearly shows the high impact of theses problem-related properties on the performance of the algorithm. In particular, two relevant hypervolume-based autocorrelation functions have been proposed for the first time, allowing to precisely estimate the ruggedness of the instance under consideration, as well as the algorithm running time.

Using the general methodology introduced in the paper applied to larger problem instances would allow to appreciate the impact of the multiobjective features on the performance of evolutionary multiobjective optimizations when tackling large-size instances. This should be possible with features that do not require the complete enumeration of the decision space, including the problem size, the number of objectives, the objective correlation, the length of a Pareto-based adaptive walk, and the hypervolume-based autocorrelation functions proposed in this paper. As well, the impact of the stopping condition, and in particular the approximation quality (the ε-value) should be carefully investigated. At last, a similar study would allow to better understand the structure of the landscape for real-world multiobjective combinatorial optimization problems. This work pushes towards the design of a *meta-algorithm* able to select the most efficient evolutionary multiobjective algorithm or parameter setting according to a prediction model based on the main problem instance features.

References

1. Aguirre, H.E., Tanaka, K.: Working principles, behavior, and performance of MOEAs on MNK-landscapes. Eur. J. Oper. Res. **181**(3), 1670–1690 (2007)
2. Daolio, F., Verel, S., Ochoa, G., Tomassini, M.: Local optima networks and the performance of iterated local search. In: Genetic and Evolutionary Computation Conference (GECCO 2012), pp. 369–376 (2012)
3. Ehrgott, M.: Multicriteria Optimization, 2nd edn. Springer, Heidelberg (2005)
4. Gorski, J., Klamroth, K., Ruzika, S.: Connectedness of efficient solutions in multiple objective combinatorial optimization. J. Optim. Theory Appl. **150**(3), 475–497 (2011)
5. Hansen, N., Auger, A., Ros, R., Finck, S., Pošík, P.: Comparing results of 31 algorithms from the black-box optimization benchmarking BBOB-2009. In: Conference on Genetic and Evolutionary Computation (GECCO 2010), pp. 1689–1696 (2010)
6. Kauffman, S.A.: The Origins of Order. Oxford University Press, New York (1993)
7. Laumanns, M., Thiele, L., Zitzler, E.: Running time analysis of evolutionary algorithms on a simplified multiobjective knapsack problem. Nat. Comput. **3**(1), 37–51 (2004)
8. Papadimitriou, C.H., Yannakakis, M.: On the approximability of trade-offs and optimal access of web sources. In: Symposium on Foundations of Computer Science (FOCS 2000), pp. 86–92. IEEE (2000)
9. Paquete, L., Camacho, C., Figueira, J.R.: A two-phase heuristic for the biobjective 0/1 knapsack problem. Unpublished (2008)
10. Paquete, L., Stützle, T.: Clusters of non-dominated solutions in multiobjective combinatorial optimization: an experimental analysis. In: Barichard, V., et al. (eds.) Multiobjective Programming and Goal Programming. Lecture Notes in Economics and Mathematical Systems, vol. 618, pp. 69–77. Springer, Heidelberg (2009)

11. Paquete, L., Schiavinotto, T., Stützle, T.: On local optima in multiobjective combinatorial optimization problems. Ann. Oper. Res. **156**(1), 83–97 (2007)
12. Verel, S., Liefooghe, A., Jourdan, L., Dhaenens, C.: On the structure of multiobjective combinatorial search space: MNK-landscapes with correlated objectives. Eur. J. Oper. Res. **227**(2), 331–342 (2013)
13. Weinberger, E.D.: Correlated and uncorrelated fitness landscapes and how to tell the difference. Biol. Cybern. **63**(5), 325–336 (1990)
14. Zitzler, E., Thiele, L., Laumanns, M., Fonseca, C.M., Grunert da Fonseca, V.: Performance assessment of multiobjective optimizers: an analysis and review. IEEE Trans. Evol. Comput. **7**(2), 117–132 (2003)

Log-log Convergence for Noisy Optimization

S. Astete-Morales$^{(\boxtimes)}$, J. Liu$^{(\boxtimes)}$, and Olivier Teytaud$^{(\boxtimes)}$

TAO (Inria), LRI, UMR 8623 (CNRS - Université Paris-Sud), Orsay, France
{sandra-cecilia.astete-morales,jialin.liu,olivier.teytaud}@inria.fr

Abstract. We consider noisy optimization problems, without the assumption of variance vanishing in the neighborhood of the optimum. We show mathematically that simple rules with exponential number of resamplings lead to a log-log convergence rate. In particular, in this case the log of the distance to the optimum is linear on the log of the number of resamplings. As well as with number of resamplings polynomial in the inverse step-size. We show empirically that this convergence rate is obtained also with polynomial number of resamplings. In this polynomial resampling setting, using classical evolution strategies and an *ad hoc* choice of the number of resamplings, we seemingly get the same rate as those obtained with specific Estimation of Distribution Algorithms designed for noisy setting.

We also experiment non-adaptive polynomial resamplings. Compared to the state of the art, our results provide (i) proofs of log-log convergence for evolution strategies (which were not covered by existing results) in the case of objective functions with quadratic expectations and constant noise, (ii) log-log rates also for objective functions with expectation $\mathbb{E}[f(x)] = ||x - x^*||^p$, where x^* represents the optimum (iii) experiments with different parameterizations than those considered in the proof. These results propose some simple revaluation schemes. This paper extends [1].

1 Introduction

In this introduction, we first present the noisy optimization setting and the local case of it. We then classify existing optimization algorithms for such settings. Afterwards we discuss log-linear and log-log scales for convergence and give an overview of the paper. In all the paper, log represents the natural logarithm and $\mathcal{N}$ is a standard Gaussian random variable (possibly multidimensional, depending on the context), except when it is specified explicitly that $\mathcal{N}$ may be any random variable with bounded density.

Noisy optimization. This term will denote the optimization of an objective function which has internal stochastic effects. When the algorithm requests $fitness(\cdot)$ of a point x, it gets in fact $fitness(x,\theta)$ for a realization of a random variable θ. All calls to $fitness(\cdot)$ are based on independent realizations of the same random variable θ. The goal of a noisy optimization algorithm is to find x such that $\mathbb{E}(fitness(x,\theta))$ is minimized (or nearly minimized).

Local noisy optimization. Local noisy optimization refers to the optimization of an objective function in which the main problem is noise, and not local

© Springer International Publishing Switzerland 2014
P. Legrand et al. (Eds.): EA 2013, LNCS 8752, pp. 16–28, 2014.
DOI: 10.1007/978-3-319-11683-9_2

minima. Hence, diversity mechanisms as in [2] or [3], in spite of their qualities, are not relevant here. We also restrict our work to noisy settings in which noise does not decrease to 0 around the optimum. This constrain makes our work different from [4]. In [5,6] we can find noise models related to ours but the results presented here are not covered by their analysis. On the other hand, in [7–9], different noise models (with Bernoulli fitness values) are considered, inclusing a noise with variance which does not decrease to 0 (as in the present paper). They provide general lower bounds, or convergence rates for specific algorithms, whereas we consider convergence rates for classical evolution strategies equipped with resamplings.

Classification of local noisy optimization algorithms. We classify noisy local convergence algorithms in the following 3 families:

- *Algorithms based on sampling, as far as they can, close to the optimum.* In this category, we include evolution strategies [5,6,10] and EDA [11] as well as pattern search methods designed for noisy cases [12–14]. Typically, these algorithms are based on noise-free algorithms, and evaluate individuals multiple times in order to cancel (reduce) the effect of noise. Authors studying such algorithms focus on the number of resamplings; it can be chosen by estimating the noise level [15], or using the step-size, or, as in parts of the present work, in a non-adaptive manner.
- *Algorithms which learn (model) the objective function,* sample at locations in which the model is not precise enough, and then assume that the optimum is nearly the optimum of the learnt model. Surrogate models and Gaussian processes [2,16] belong to this family. However, Gaussian processes are usually supposed to achieve global convergence (i.e. good properties on multimodal functions) rather than local convergence (i.e. good properties on unimodal functions) - in this paper, we focus on local convergence.
- *Algorithms which combine both ideas,* assuming that learning the objective function is a good idea for handling noise issues but considering that points too far from the optimum cannot be that useful for an optimization. This assumption makes sense at least in a scenario in which the objective function cannot be that easy to learn on the whole search domain. CLOP [7,8] is such an approach.

Log-linear scale and log-log scale: uniform and non-uniform rates. To ensure the convergence of an algorithm and analyze the rate at which it converges are part of the main goals when it comes to the study of optimization algorithms.

In the noise-free case, evolution strategies typically converge linearly in log-linear scale, this is, the logarithm of the distance to the optimum typically scales linearly with the number of evaluations (see Sect. 2.1 for more details on this). The case of noisy fitness values leads to a log-log convergence [9]. We investigate conditions under which such a log-log convergence is possible. In particular, we focus on uniform rates. Uniform means that all points are under a linear curve in the log-log scale. Formally, the rate is the infimum of C such that with probability $1 - \delta$, for m sufficiently large, all iterates after m fitness evaluations verify $\log ||x_m|| \leq -C \log m$, where x_m is the mth evaluated individual. This is, all points are supposed to be "good" (i.e. satisfy the inequality); not only the best point of a given iteration. In contrast, a non-uniform rate would be the infimum of C such that $\log ||x_{k_m}|| \leq -C \log k_m$ for some increasing sequence k_m.

The state of the art in this matter exhibits various results. For an objective function with expectation $\mathbb{E}[f(x)] = ||x-x^*||^2$, when the variance is not supposed to decrease in the neighborhood of the optimum, it is known that the best possible slope in this log-log graph is $-\frac{1}{2}$ (see [17]), but without uniform rate. When optimizing $f(x) = ||x||^p + \mathcal{N}$, this slope is provably limited to $-\frac{1}{p}$ under locality assumption (i.e. when sampling far from the optimum does not help, see [9] for a formalization of this assumption), and it is known that some ad hoc EDA can reach $-\frac{1}{2p}$ (see [18]).

For evolution strategies, the slope is not known. Also, the optimal rate for $\mathbb{E}[f(x)] = ||x-x^*||^p$ for $p \neq 2$ is unknown; we show that our evolution strategies with simple revaluation schemes have linear convergence in log-log representation in such a case.

Algorithms considered in this paper. We here focus on simple revaluation rules in evolution strategies, based on choosing the number of resamplings. We start with rules which decide the number of revaluations only depending on the iteration number n. This is, independently of the step-size σ_n, the parents x_n and fitness values. To the best of our knowledge, these simple rules have not been analyzed so far. Nonetheless, they have strong advantages: we get a linear slope in log-log curve simple rules only depending on n whereas rules based on numbers of resamplings defined as a function of σ_n have a strong sensitivity to parameters. Also evolution strategies, contrarily to algorithms with good non-uniform rates, have a nice empirical behavior from the point of view of uniform rates, as shown mathematically by [18].

Overview of the paper. In this paper we show mathematical proofs and experimental results on the convergence of the evolutionary algorithms that will be described in the following sections, which include some resampling rules aiming to cancel the effect of noise. The theoretical analysis presents an exponential number of resamplings together with an assumption of scale invariance. This result is extended to an adaptive rule of resamplings (Sect. 2.3), in which the number of evaluations depend on the step size only; we also get rid of the scale invariant assumption. Essentially, the algorithms for which we get a proof have the same dynamics as in the noise-free case, they just use enough resamplings for cancelling the noise. This is consistent with the existing literature, in particular [18] which shows a log-log convergence for an Estimation of Distribution Algorithm with exponentially decreasing step-size and exponentially increasing number of resamplings.

In the experimental part, we see that another solution is a polynomially increasing number of resamplings (independently of σ_n; the number of resamplings just smoothly increases with the number of iterations, in a non-adaptive manner), leading to a slower convergence when considering the progress rate per iteration, but the same log-log convergence when considering the progress rate per evaluation. We could get positive experimental results even with the non-proved polynomial number of revaluations (non-adaptive); maybe those results are the most satisfactory (stable) results. We could also get convergence with adaptive rules (number of resamplings depending on the step-size), however results are seemingly less stable than with non-adaptive methods.

2 Theoretical Analysis: Exponential Non-adaptive Rules Can Lead to Log/log Convergence

Section 2.1 is devoted to some preliminaries. Section 2.2 presents results in the scale invariant case, for an exponential number of resamplings and non-adaptive rules. Section 2.3 will focus on adaptive rules, with numbers of resamplings depending on the step-size.

2.1 Preliminary: Noise-Free Case

In the noise-free case, for some evolution strategies, we know the following results, almost surely (see e.g. Theorem 4 in [19], where, however, the negativity of the constant is not proved and only checked by Monte-Carlo simulations): $\log(\sigma_n)/n$ converges to some constant $(-A) < 0$ and $\log(||x_n||)/n$ converges to some constant $(-A') < 0$.

This implies that for any $\rho < A$, $\log(\sigma_n) \leq -\rho n$ for n sufficiently large. So, $\sup_{n \geq 1} \log(\sigma_n) + \rho n$ is finite. With these almost sure results, now consider V the quantile $1 - \delta/4$ of $\exp\left(\sup_{n \geq 1} \log(\sigma_n) + \rho n\right)$. Then, with probability at least $1 - \delta/4$, $\forall n \geq 1, \sigma_n \leq V \exp(-\rho n)$. We can apply the same trick for lower bounding σ_n, and upper and lower bounding $||x_n||$, all of them with probability $1 - \delta/4$, so that all bounds hold true simultaneously with probability at least $1 - \delta$.

Hence, for any $\alpha < A'$, $\alpha' > A'$, $\rho < A$, $\rho' > A$, there exist $C > 0$, $C' > 0$, $V > 0$, $V' > 0$, such that with probability at least $1 - \delta$

$$\forall n \geq 1, \quad C' \exp(-\alpha' n) \leq ||x_n|| \leq C \exp(-\alpha n); \tag{1}$$

$$\forall n \geq 1, \quad V' \exp(-\rho' n) \leq \sigma_n \leq V \exp(-\rho n). \tag{2}$$

We will first show, in Sect. 2.2, our noisy optimization result (Theorem 1):

(i) in the scale invariant case
(ii) using Eq. 1 (supposed to hold in the noise-free case).

We will then show similar results in Sect. 2.3:

(i) without scale-invariance
(ii) using Eq. 2 (supposed to hold in the noise-free case)
(iii) with other resamplings schemes.

2.2   Scale Invariant Case, with Exponential Number of Resamplings

We consider Algorithm 1, a version of multi-membered Evolution Strategies, the (μ,λ)-ES. μ denotes the number of parents and λ the number of offspring ($\mu \leq \lambda$). In every generation, the selection takes place among the λ offspring, produced from a population of μ parents. Selection is based on the ranking of the individuals according their $fitness(\cdot)$ taking the μ best individuals among the population. Here x_n denotes the parent at iteration n.

Algorithm 1. An evolution strategy, with exponential number of resamplings. If we consider $K = 1$ and $\zeta = 1$ we obtain the case without resampling. $\mathcal{N}$ is an arbitrary random variable with bounded density (each use is independent of others).

Parameters: $K > 0$, $\zeta \geq 0$, $\lambda \geq \mu > 0$, a dimension $d > 0$.
Input: an initial $x_1 \in \mathbb{R}^d$ and an initial $\sigma_0 > 0$.
$n \leftarrow 1$
while (true) **do**
    Generate λ individuals $i_1, \ldots, i_\lambda$ independently using
$$i_j = x_n + \sigma_{n,j}\mathcal{N}. \qquad (3)$$
    Evaluate each of them $r_n = \lceil K\zeta^n \rceil$ times and average their fitness values.
    Select the μ best individuals $j_1, \ldots, j_\mu$.
    Update: from x, σ_n, $i_1, \ldots, i_\lambda$ and $j_1, \ldots, j_\mu$, compute x_{n+1} and σ_{n+1}.
    $n \leftarrow n+1$
end while

We now state our first theorem, under log-linear convergence assumption (the assumption in Eq. 5 is just Eq. 1).

Theorem 1. *Consider the fitness function*

$$f(z) = ||z||^p + \mathcal{N} \qquad (4)$$

over $\mathbb{R}^d$ and $x_1 = (1, 1, \ldots, 1)$.
Consider an evolution strategy with population size λ, parent population size μ, such that without resampling, for any $\delta > 0$, for some $\alpha > 0$, $\alpha' > 0$, with probability $1 - \delta/2$, with objective function $fitness(x) = ||x||$,

$$\exists C, C'; \qquad C' \exp(-\alpha'n) \leq ||x_n|| \leq C \exp(-\alpha n). \qquad (5)$$

Assume, additionally, that there is scale invariance:

$$\sigma_n = C''||x_n|| \qquad (6)$$

for some $C'' > 0$.
Then, for any $\delta > 0$, there is $K_0 > 0, \zeta_0 > 0$ such that for $K \geq K_0$, $\zeta > \zeta_0$, Eq. 1 also holds with probability at least $1 - \delta$ for fitness function as in Eq. 4 and resampling rule as in Algorithm 1.

Remarks: *(i) Informally speaking, our theorem shows that if a scale invariant algorithm converges in the noise-free case, then it also converges in the noisy case with the exponential resampling rule, at least if parameters are large enough (a similar effect of constants was pointed out in [4] in a different setting).*

(ii) We assume that the optimum is in 0 and the initial x_1 at 1. Note that these assumptions have no influence when we use algorithms invariant by rotation and translation.

(iii) We show a log-linear convergence rate as in the noise-free case, but at the cost of more evaluations per iteration. When normalized by the number of function evaluations, we get $\log \|x_n\|$ linear in the logarithm of the number of function evaluations, as detailed in Corollary 1.

Proof of the theorem: In all the proof, $\mathcal{N}$ denotes a standard Gaussian random variable (depending on the context, in dimension 1 or d). Consider an arbitrary $\delta > 0$, $n \geq 1$ and $\delta_n = \exp(-\gamma n)$ for some $\gamma > 0$.

Define p_n the probability that two generated points, e.g. i_1 and i_2, are such that $|\ \|i_1\|^p - \|i_2\|^p\ | \leq \delta_n$.

Step 1: Using Eqs. 3 and 6, we show that

$$p_n \leq B' \exp(-\gamma' n) \tag{7}$$

for some $B' > 0, \gamma' > 0$ depending on γ, d, p, C', C'', α'.

Proof of step 1: with $\mathcal{N}_1$ and $\mathcal{N}_2$ two d-dimensional independent standard Gaussian random variables,

$$p_n \leq P(|\ \|1 + C''\mathcal{N}_1\|^p - \|1 + C''\mathcal{N}_2\|^p\ | \leq \delta_n/\|x_n\|^p). \tag{8}$$

Define $densityMax$ the supremum of the density of $|\ \|1+C''\mathcal{N}_1\|^p - \|1+C''\mathcal{N}_2\|^p\ |$ we get

$$p_n \leq densityMax\, C'^{-p} \exp((p\alpha' - \gamma)n),$$

hence the expected result with $\gamma' = \gamma - p\alpha'$ and $B' = densityMax(C')^{-p}$. Notice that $densityMax$ is upper bounded.

In particular, γ' is arbitrarily large, provided that γ is sufficiently large.

Step 2: Consider now $p_n^{(1)}$ the probability that there exists i_1 and i_2 such that $|\ \|i_1\|^p - \|i_2\|^p\ | \leq \delta_n$. Then, $p_n^{(1)} \leq \lambda^2 p_n \leq B'\lambda^2 \exp(-\gamma' n)$.

Step 3: Consider now $p_n^{(2)}$ the probability that $|\mathcal{N}/\sqrt{K\zeta^n}| \geq \delta_n/2$. First, we write $p_n^{(2)} = P(\mathcal{N} \geq \frac{\delta_n}{2}\sqrt{K\zeta^n})$. So by Chebychev inequality, $p_n^{(2)} \leq B'' \exp(-\gamma'' n)$ for $\gamma'' = \log(\zeta) - 2\gamma$ arbitrarily large, provided that ζ is large enough, and $B'' = 4/K$.

Step 4: Consider now $p_n^{(3)}$ the probability that $|\mathcal{N}/\sqrt{K\zeta^n}| \geq \delta_n/2$ at least once for the λ evaluated individuals of iteration n. Then, $p_n^{(3)} \leq \lambda p_n^{(2)}$.

Step 5: In this step we consider the probability that two individuals are mis-ranked due to noise. Let us now consider $p_n^{(4)}$ the probability that at least two points i_a and i_b at iteration n verify

$$||i_a||^p \leq ||i_b||^p \tag{9}$$

$$\text{and} \quad noisyEvaluation(i_a) \geq noisyEvaluation(i_b) \tag{10}$$

where $noisyEvaluation(i)$ is the average of the multiple evaluations of individual i. Equations 9 and 10 occur simultaneously if either two points have very similar fitness (difference less than δ_n) or the noise is big (larger than $\delta_n/2$). Therefore, $p_n^{(4)} \leq p_n^{(1)} + p_n^{(3)} \leq \lambda^2 p_n + \lambda p_n^{(2)}$ so $p_n^{(4)} \leq (B' + B'')\lambda^2 \exp(-\min(\gamma', \gamma'')n)$.

Step 6: Step 5 was about the probability that at least two points at iteration n are misranked due to noise. We now consider $\sum_{n \geq 1} p_n^{(4)}$, which is an upper bound on the probability that in at least one iteration there is a misranking of two individuals.

If γ' and γ'' are large enough, $\sum_{n \geq 1} p_n^{(4)} < \delta$.

This implies that with probability at least $1 - \delta$, provided that K and ζ have been chosen large enough for γ and γ' to be large enough, we get the same rankings of points as in the noise free case - this proves the expected result. $\square$

The following corollary shows that this is a log-log convergence.

Corollary 1: log-log **convergence with exponential resampling.** *With e_n the number of evaluations at the end of iteration n, we have $e_n = K\zeta\frac{\zeta^n - 1}{\zeta - 1}$. We then get, from Eq. 1,*

$$\log(||x_n||)/\log(e_n) \rightarrow -\frac{\alpha}{\log \zeta} \tag{11}$$

with *probability at least $1 - \delta$. Eq. 11 is the convergence in* log/log *scale.*

We have shown this property for an exponentially increasing number of resamplings, which is indeed similar to R-EDA [18], which converges with a small number of iterations but with exponentially many resamplings per iteration. In the experimental Sect. 3, we will check what happens in the polynomial case.

2.3 Extension: Adaptive Resamplings and Removing the Scale Invariance Assumption

We have assumed above a scale invariance. This is obviously not a nice feature of our proof, because scale invariance does not correspond to anything real; in a real setting we do not know the distance to the optimum. We show below an extension of the result above using the assumption of a log-linear convergence of σ_n as in Eq. 2 instead of the scale invariance used before.

In the corollary below, we also get rid of the non-adaptive rule with exponential number of resamplings, replaced by a number of resamplings depending on the step-size σ_n only, as in Eq. 2. In one corollary, we switch to both (i) adaptive resampling rule and (ii) no scale invariance; each change can indeed be proved independently of the other.

Algorithm 2. An evolution strategy, with number of resamplings polynomial in the step-size. The case without resampling means $Y = 1$ and $\eta = 0$. $\mathcal{N}$ is an arbitrary random variable with bounded density (each use is independent of others).

Parameters: $Y > 0, \eta \geq 0$, $\lambda \geq \mu > 0$, a dimension $d > 0$.
Input: an initial $x_1 \in \mathbb{R}^d$ and an initial $\sigma_0 > 0$.
$n \leftarrow 1$
while (true) **do**
    Generate λ individuals $i_1, \ldots, i_\lambda$ independently using
$$i_j = x_n + \sigma_{n,j}\mathcal{N}. \tag{12}$$

    Evaluate each of them $r_n = \lceil Y\sigma_n^{-\eta}\rceil$ times and average their fitness values.
    Select the μ best individuals $j_1, \ldots, j_\mu$.
    Update: from x, σ_n, $i_1, \ldots, i_\lambda$ and $j_1, \ldots, j_\mu$, compute x_{n+1} and σ_{n+1}.
    $n \leftarrow n + 1$
end while

Corollary 2: adaptive resampling and no scale-invariance. *The proof of Theorem [1] also holds without scale invariance, under the following assumptions:*

- *For any $\delta > 0$, there are constants $\rho > 0, V > 0, \rho' > 0, V' > 0$ such that with probability at least $1 - \delta$, Eq. 2 holds.*
- *The number of revaluations is*

$$Y\left(\frac{1}{\sigma_n}\right)^\eta \tag{13}$$

with Y and η sufficiently large.
- *Individuals are still randomly drawn using $x_n + \sigma_n\mathcal{N}$ for some random variable $\mathcal{N}$ with bounded density.*

Remark: *This setting is useful in cases like self-adaptive algorithms, in which we do not use directly a Gaussian random variable, but a Gaussian random variable multiplied e.g. by $\exp(\frac{1}{\sqrt{d}})$Gaussian, with Gaussian a standard Gaussian random variable. For example, SA-ES algorithms as in [19] are included in this proof because they converge log-linearly as explained in Sect. 2.1.*

Proof of corollary 2: Two steps of the proof are different, namely step 1 and step 2. We here adapt the proofs of these two steps.

Adapting step 1: Eq. 8 becomes Eq. 14:

$$p_n \leq P(|\;||1 + C_n''\mathcal{N}_1||^p - ||1 + C_n''\mathcal{N}_2||^p\;| \leq \delta_n/||x_n||^p). \tag{14}$$

where $C_n'' = \sigma_n/||x_n|| \geq t'\exp(-tn)$ for some $t > 0, t' > 0$ depending on ρ, ρ', V, V' only. Equation 14 leads to

$$p_n \leq (C_n'')^{-d} densityMaxC'^{-p}\exp((p\alpha' - \gamma)n),$$

hence the expected result with $\gamma' = \gamma - p\alpha' - dt$. $densityMax$ is upper bounded due to the third condition of corollary 2.

Adapting step 2: It is sufficient to show that the number of resamplings is larger (for each iteration) than in the Theorem 1, so that step 2 still holds.

Equation 13 implies that the number of revaluations at step n is at least $Y\left(\frac{1}{V}\right)^{\eta} \exp(\rho\eta n)$. This is more than $K\zeta^n$, at least if Y and η are large enough. This leads to the same conclusion as in the Theorem 1, except that we have probability $1 - 2\delta$ instead of $1 - \delta$ (which is not a big issue as we can do the same with $\delta/2$). □

The following corollary is here for showing that our result leads to the log-log convergence.

Corollary 3: log-log **convergence for adaptive resampling.** *With e_n the number of evaluations at the end of iteration n, we have* $e_n = Y\left(\frac{1}{V}\right)^{\eta}$ $\exp(\rho\eta)\frac{\exp(\rho\eta n)-1}{\exp(\rho\eta)-1}$. *We then get, from Eq. 1,*

$$\log(||x_n||)/\log(e_n) \to -\frac{\alpha}{\rho\eta} \tag{15}$$

with probability at least $1 - \delta$. Equation 15 is the convergence in log/log *scale.*

3 Polynomial Number of Resamplings: Experiments

We here consider a polynomial number of resamplings, as in Algorithm 3.

Algorithm 3. An evolution strategy, with polynomial number of resamplings. The case without resampling means $K = 1$ and $\zeta = 0$.

Parameters: $K > 0$, $\zeta \geq 0$, $\lambda \geq \mu > 0$, a dimension $d > 0$.
Input: an initial $x_1 \in \mathbb{R}^d$ and an initial $\sigma_0 > 0$.
$n \leftarrow 1$
while (true) **do**
  Generate λ individuals $i_1, \ldots, i_\lambda$ independently using

$$\sigma_{n,j} = \sigma_n \times \exp(\frac{1}{\sqrt{d}}\mathcal{N}) \tag{16}$$

$$i_j = x_n + \sigma_{n,j}\mathcal{N}.$$

  Evaluate each of them $r_n = \lceil Kn^\zeta \rceil$ times and average their fitness values.
  Select the μ best individuals $j_1, \ldots, j_\mu$.
  Update: from x, σ_n, $i_1, \ldots, i_\lambda$ and $j_1, \ldots, j_\mu$, compute x_{n+1} and σ_{n+1}.
  $n \leftarrow n + 1$
end while

Experiments are performed in a "real" setting, without scale invariance. Importantly, our mathematical results hold only log-log convergence under the

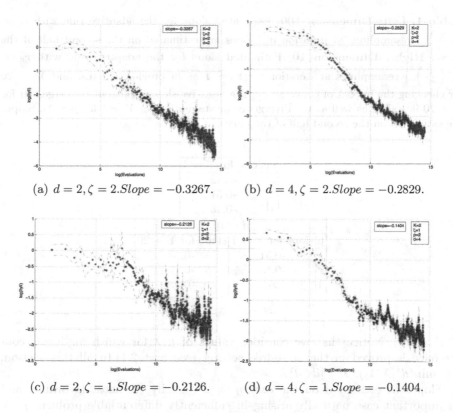

(a) $d = 2, \zeta = 2. Slope = -0.3267.$ (b) $d = 4, \zeta = 2. Slope = -0.2829.$

(c) $d = 2, \zeta = 1. Slope = -0.2126.$ (d) $d = 4, \zeta = 1. Slope = -0.1404.$

Fig. 1. Experiments in dimension 2, 3, 4 with $\zeta = 1, 2$ (number of evaluations shown by x-axis) for $r_n = K\lceil n^\zeta \rceil$ (i.e. polynomial, non-adaptive) with $\mu = 2$, $\lambda = 4$, $p = 2$ and $K = 2$. The slope is evaluated on the second half of the iterations. We get slopes close to $-1/(2p)$. All results are averaged over 20 runs.

assumption that constants are large enough. We present results with fitness function $f(x) = ||x||^p + \mathcal{N}$ with $p = 2$ in Fig. 1.

In experiments with the following parameters (as recommended in [10,20]): $p = 1$ or $p = 4$, dimension 2, 3, 4, 5, $\zeta = 1, 2, 3$, $\mu = \min(d, \lceil \lambda/4 \rceil)$, $\lambda = \lceil d\sqrt{d} \rceil$, slopes are usually better than $-1/(2p)$ for $\zeta = 2$ or $\zeta = 3$ and worse for $\zeta = 1$. Non-presented experiments show that $\zeta = 0$ performs very poorly. Seemingly results for ζ large are farther from the asymptotic regime. We conjecture that the asymptotic regime is $-1/(2p)$ but that it is reached later when ζ is large. R-EDA [18] reaches $-1/(2p)$; we seemingly get slightly better but this might be due to a non-asymptotic effect. Figure 1 provides results with high numbers of evaluations.

4 Experiments with Adaptivity: $Y\sigma_n^{-\eta}$ Revaluations

We here show experiments with Algorithm 2. The algorithm should converge linearly in log-log scale as shown by Corollary 3, at least for large enough values

Table 1. Left: Dimension 100. Estimated slope for the adaptive rule with $r_n = \lceil \left(\frac{1}{\sigma_n}\right)^2 \rceil$ resamplings at iteration n. Slopes are estimated on the second half of the curve. **Right: Dimension 10.** Estimated slope for the adaptive rule with $r_n = \lceil Y \left(\frac{1}{\sigma_n}\right)^2 \rceil$ resamplings at iteration n ($Y = 1$ as in previous curves, and $Y = 20$ for checking the impact of convergence; the negative slope (apparent convergence) for $Y = 20$ is stable, as well as the divergence or stagnation for $Y = 1$ for $p = 4$). Slopes are estimated on the second half of the curve.

	$d = 100$
p	slope for $Y = 1$
1	-0.52
2	-0.51
4	-0.45

	$d = 10$	
p	slope for $Y = 1$	slope for $Y = 20$
1	-0.51	-0.50
2	-0.18	-0.17
4	>0	-0.08

of Y and η. Notice that we consider values of μ, λ for which log-linear convergence is proved in the noise-free setting (see Sect. 2.1).In all this section, $\mu = \min(d, \lceil \lambda/4 \rceil)$, $\lambda = \lceil d\sqrt{d} \rceil$.

Slopes as estimated on the case $\eta = 2$ (usually the most favorable, and an important case naturally arising in sufficiently differentiable problems) are given in Table 1 (left) for dimension $d = 100$. In this case we are far from the asymptotic regime.

We get results close to $-\frac{1}{2}$ in all cases This slope of $-\frac{1}{2}$ is reachable by algorithms which learn a model of the fitness function, as e.g. [7]. In this case of high dimension we are far from the slope $1/(-2p)$, which might be the case for the asymptotic results. This is suggested by experiments in dimension 10 summarized in Table 1 (right). We also point out that the known complexity bounds is $-\frac{1}{p}$ (from [9]), and maybe the slope can reach $-\frac{1}{p}$ in some cases.

Results with $Y \left(\frac{1}{\sigma}\right)^\eta$ are moderately stable (impact of Y, in particular). This supports our preference for stable rules, such as non-adaptively choosing n^2 revaluations per individual at iteration n.

5   Conclusion

We have shown mathematically log-log convergence results and studied experimentally the slope in this convergence. These results were shown for evolution strategies, which are known for having good uniform rates, rather than good non-uniform rates. We summarize these two parts below and give some research directions.

Log-log convergence. We have shown that the log-log convergence (i.e. linear convergence with x-axis the log of the number of evaluations and y-axis the log of the distance to the optimum) occurs in various cases:

- Non-adaptive rules, with number of resamplings exponential in the iteration counter. Here we have a mathematical proof, which includes the assumption of scale invariance; as shown by Corollary 2, this can be extended to non scale-invariant algorithms;
- Adaptive rules, with number of resamplings polynomial in $1/\sigma_n$ with σ_n the step-size. Here we have a mathematical proof; however, there is a strong sensitivity to constants Y and η which participate in the number of resamplings per individual, $Y\left(\frac{1}{\sigma_n}\right)^{\eta}$;
- Non-adaptive rule, with polynomial number of resamplings. This case is a quite convenient scheme experimentally but we have no proof.

Slope in log-log convergence. Experimentally, the best slope in the log-log representation is often close to $-\frac{1}{2p}$ for fitness function $f(x) = ||x||^p + \mathcal{N}$. It is known that under modeling assumptions (i.e. the function is regular enough for being optimized by learning), it is possible to do better than that (the slope becomes $-1/2$ for parametric cases, see [7] and references therein), but $-\frac{1}{2p}$ is the best known exponent under locality assumption. Basically, locality assumption ensures that most points are reasonably good, whereas some specialized noisy optimization algorithms sample a few very good points and essentially sample individuals far from the optimum (see e.g. [7]).

Further work. The main further work is the mathematical analysis of the polynomial number of resamplings in the non-adaptive case. Also, a combination of adaptive and non-adaptive rules might be interesting; adaptive rules are intuitively satisfactory, but non-adaptive polynomial rules provide simple efficient solutions, with empirically easy (no tuning) results. If our life depended on a scheme, we would for the moment choose a simple polynomial rule with a number of revaluations quadratic in the number of evaluations, in spite of (maybe) moderate elegance due to lack of adaptivity.

References

1. Morales, S.A., Liu, J., Teytaud, O.: Noisy optimization convergence rates. In: GECCO (Companion), pp. 223–224 (2013)
2. Jones, D., Schonlau, M., Welch, W.: Efficient global optimization of expensive black-box functions. J. Global Optim. **13**(4), 455–492 (1998)
3. Auger, A., Jebalia, M., Teytaud, O.: Algorithms (X, sigma, eta): quasi-random mutations for evolution strategies. In: Talbi, E.-G., Liardet, P., Collet, P., Lutton, E., Schoenauer, M. (eds.) EA 2005. LNCS, vol. 3871, pp. 296–307. Springer, Heidelberg (2006)
4. Jebalia, M., Auger, A., Hansen, N.: Log linear convergence and divergence of the scale-invariant (1+1)-ES in noisy environments. Algorithmica **59**(3), 425–460 (2010)

5. Arnold, D.V., Beyer, H.G.: A general noise model and its effects on evolution strategy performance. IEEE Trans. Evol. Comput. **10**, 380–391 (2006)
6. Finck, S., Beyer, H.G., Melkozerov, A.: Noisy optimization: a theoretical strategy comparison of ES, EGS, SPSA & IF on the noisy sphere. In: GECCO, pp. 813–820 (2011)
7. Coulom, R.: CLOP: confident local optimization for noisy Black-Box parameter tuning. In: van den Herik, H.J., Plaat, A. (eds.) ACG 2011. LNCS, vol. 7168, pp. 146–157. Springer, Heidelberg (2012)
8. Coulom, R., Rolet, P., Sokolovska, N., Teytaud, O.: Handling expensive optimization with large noise. In: Foundations of Genetic Algorithms (2011)
9. Teytaud, O., Decock, J.: Noisy optimization complexity. In: FOGA - Foundations of Genetic Algorithms XII - 2013, Adelaide, Australie (2013)
10. Beyer, H.G.: The Theory of Evolution Strategies. Natural Computing Series. Springer, Heideberg (2001)
11. Yang, X., Birkfellner, W., Niederer, P.: Optimized 2d/3d medical image registration using the estimation of multivariate normal algorithm (EMNA). In: Biomedical Engineering (2005)
12. Anderson, E.J., Ferris, M.C.: A direct search algorithm for optimization with noisy function evaluations. SIAM J. Optim. **11**, 837–857 (2001)
13. Lucidi, S., Sciandrone, M.: A derivative-free algorithm for bound constrained optimization. Comp. Opt. Appl. **21**, 119–142 (2002)
14. Kim, S., Zhang, D.: Convergence properties of direct search methods for stochastic optimization. In: Proceedings of the Winter Simulation Conference, WSC '10, pp. 1003–1011 (2010)
15. Hansen, N., Niederberger, S., Guzzella, L., Koumoutsakos, P.: A method for handling uncertainty in evolutionary optimization with an application to feedback control of combustion. IEEE Trans. Evol. Comput. **13**, 180–197 (2009)
16. Villemonteix, J., Vazquez, E., Walter, E.: An informational approach to the global optimization of expensive-to-evaluate functions. J. Global Optim. **44**, 509–534 (2008)
17. Fabian, V.: Stochastic approximation of minima with improved asymptotic speed. Ann. Math. Stat. **38**, 191–200 (1967)
18. Rolet, P., Teytaud, O.: Bandit-based estimation of distribution algorithms for noisy optimization: rigorous runtime analysis. In: Proceedings of Lion4 (accepted); presented in TRSH 2009 in Birmingham, pp. 97–110 (2009)
19. Auger, A.: Convergence results for $(1,\lambda)$-SA-ES using the theory of φ-irreducible Markov chains
20. Fournier, H., Teytaud, O.: Lower bounds for comparison based evolution strategies using VC-dimension and sign patterns. Algorithmica **59**, 387–408 (2010)

Preventing Premature Convergence and Proving the Optimality in Evolutionary Algorithms

Charlie Vanaret[1,2](✉), Jean-Baptiste Gotteland[1,2], Nicolas Durand[1,2], and Jean-Marc Alliot[2]

[1] Laboratoire de Mathématiques Appliquées,
Informatique et Automatique pour l'Aérien, Ecole Nationale de l'Aviation Civile,
Toulouse, France
{vanaret,gottelan,durand}@recherche.enac.fr
[2] Institut de Recherche en Informatique de Toulouse, Toulouse, France
jean-marc.alliot@irit.fr

Abstract. Evolutionary Algorithms (EA) usually carry out an efficient exploration of the search-space, but get often trapped in local minima and do not prove the optimality of the solution. Interval-based techniques, on the other hand, yield a numerical proof of optimality of the solution. However, they may fail to converge within a reasonable time due to their exponential complexity and their inability to quickly compute a good approximation of the global minimum. The contribution of this paper is a hybrid algorithm called Charibde in which a particular EA, Differential Evolution, cooperates with a branch and bound algorithm endowed with interval propagation techniques. It prevents premature convergence toward local optima and is highly competitive with both deterministic and stochastic existing approaches. We demonstrate its efficiency on a benchmark of highly multimodal problems, for which we provide previously unknown global minima and certification of optimality.

1 Motivation

Evolutionary Algorithms (EA) have been widely used by the global optimization community for their ability to handle complex problems with no assumption on continuity or differentiability. They generally converge toward satisfactory solutions, but may get trapped in local optima and provide suboptimal solutions. Moreover, their convergence remains hard to control due to their stochastic nature. On the other hand, exhaustive Branch and Bound methods based on Interval Analysis [1] guarantee rigorous bounds on the solutions to numerical optimization problems but are limited by their exponential complexity.

Few methods attempted to hybridize EA and branch and bound algorithms in which lower bounds of the objective function are computed using Interval Analysis. The approaches in the literature are essentially *integrative*, in that they embed one algorithm within the other. Sotiropoulos et al. [2] used an Interval Branch and Bound (IB&B) to reduce the domain to a list of ε-large subspaces. A Genetic Algorithm (GA) [3] was then initialized within each subspace to

© Springer International Publishing Switzerland 2014
P. Legrand et al. (Eds.): EA 2013, LNCS 8752, pp. 29–40, 2014.
DOI: 10.1007/978-3-319-11683-9_3

improve the upper bound of the global minimum. Zhang and Liu [4] and Lei and Chen [5] used respectively a GA and mind evolutionary computation within the IB&B to improve the bounds and the exploration of the remaining subspaces. In a previous communication [6], we proposed a *cooperative* approach combining the efficiency of a GA and the reliability of Interval Analysis. We presented new optimality results for two multimodal benchmark functions (Michalewicz, dimension 12 and rotated Griewank, dimension 8), demonstrating the validity of the approach. However, techniques that exploit the analytical form of the objective function, such as local monotonicity and constraint programming, were not addressed. In this paper, we propose an advanced cooperative algorithm, Charibde (Cooperative Hybrid Algorithm using Reliable Interval-Based methods and Differential Evolution), in which a Differential Evolution algorithm cooperates with interval propagation methods. New optimal results achieved on a benchmark of difficult multimodal functions attest the substantial gain in performance.

The rest of the paper is organized as follows. Notations of Interval Analysis are introduced in Sect. 2 and interval-based techniques are presented in Sect. 3. The implementation of Charibde is detailed in Sect. 4. Results on a benchmark of test functions are discussed in Sect. 5.

2   Interval Analysis

Interval Analysis (IA) bounds round-off errors due to the use of floating-point arithmetic by computing interval operations with outward rounding [1]. Interval arithmetic extends real-valued functions to intervals.

Definition 1 (Notations). *An interval $X = [\underline{X}, \overline{X}]$ with floating-point bounds defines the set $\{x \in \mathbb{R} \mid \underline{X} \leq x \leq \overline{X}\}$. $\mathbb{IR}$ denotes the set of real intervals. We note $m(X) = \frac{1}{2}(\underline{X} + \overline{X})$ its midpoint. A box $\mathbf{X} = (X_1, \ldots, X_n)$ is an interval vector. We note $m(\mathbf{X}) = (m(X_1), \ldots, m(X_n))$ its midpoint. We note $\square(X, Y)$ the convex hull of two boxes X and Y, that is the smallest box that contains X and Y.*

In the following, capital letters represent interval quantities (interval X) and bold letters represent vectors (box $\mathbf{X}$, vector $\mathbf{x}$).

Definition 2 (Interval extension; Natural interval extension). *Let $f : \mathbb{R}^n \to \mathbb{R}$ be a real-valued function. $F : \mathbb{IR}^n \to \mathbb{IR}$ is an interval extension of f if*

$$\forall \mathbf{X} \in \mathbb{IR}^n, f(\mathbf{X}) = \{f(\mathbf{x}) \mid \mathbf{x} \in \mathbf{X}\} \subset F(\mathbf{X})$$
$$\forall (\mathbf{X}, \mathbf{Y}) \in \mathbb{IR}^n, \mathbf{X} \subset \mathbf{Y} \Rightarrow F(\mathbf{X}) \subset F(\mathbf{Y})$$

The natural interval extension F_N is obtained by replacing the variables with their domains and real elementary operations with interval arithmetic operations.

The quality of enclosure of $f(X)$ depends on the syntactic form of f: the natural interval extensions of different but equivalent expressions may yield different

ranges (Example 1). In particular, IA generally computes a large overestimation of the image due to multiple occurrences of a same variable, considered as
different variables. This **dependency problem** is the main source of overestimation when using interval computations. However, an appropriate rewriting
of the expression may reduce or overcome dependency: if f is continuous inside
a box, its natural interval extension F_N yields the optimal image when each
variable occurs only once in its expression.

Example 1. Let $f(x) = x^2 - 2x, g(x) = x(x - 2)$ and $h(x) = (x - 1)^2 - 1$, where
$x \in X = [1, 4]$. f, g and h have equivalent expressions, however computing their
natural interval extensions yields

$$F_N([1, 4]) = [1, 4]^2 - 2 \times [1, 4] = [1, 16] - [2, 8] = [-7, 14]$$
$$G_N([1, 4]) = [1, 4] \times ([1, 4] - 2) = [1, 4] \times [-1, 2] = [-4, 8]$$
$$H_N([1, 4]) = ([1, 4] - 1)^2 - 1 = [0, 3]^2 - 1 = [0, 9] - 1 = [-1, 8]$$

We have $f([1, 4]) = H_N([1, 4]) \subset G_N([1, 4]) \subset F_N([1, 4])$.

3   Interval-Based Techniques

Interval Branch and Bound Algorithms (IB&B) exploit the conservative
properties of interval extensions to rigorously bound global optima of numerical
optimization problems [7]. The method consists in splitting the initial search-
space into subspaces (branching) on which an interval extension F of the objective function f is evaluated (bounding). By keeping track of the best upper bound
$\tilde{f}$ of the global minimum f^*, boxes that certainly do not contain a global minimizer are discarded (Example 2). The remaining boxes are stored to be processed
at a later stage until the desired precision ε is reached. The process is repeated
until all boxes have been processed. Convergence certifies that $\tilde{f} - f^* < \varepsilon$, even
in the presence of rounding errors. However, the exponential complexity of IB&B
hinders the speed of convergence on large problems.

Example 2. Let us detail the first step of the IB&B on the problem $\min\limits_{x \in X} f(x) = x^4 - 4x^2$ over the interval $X = [-1, 4]$. The natural interval extension of f is
$F_N(X) = X^4 - 4X^2$ and $F_N([-1, 4]) = [-64, 256] \supset [-4, 192] = f([-1, 4])$. The
floating-point evaluation $f(1) = -3$ yields an upper bound $\tilde{f}$ of f^*. Evaluating
F_N on the subinterval $[3, 4]$ reduces the overestimation induced by the dependency effect: $F_N([3, 4]) = [17, 220] \supset [45, 192] = f([3, 4])$. Since this enclosure
is rigorous, $\forall x \in [3, 4], f(x) \geq 17 > \tilde{f} = -3 \geq f^*$. Therefore, the interval $[3, 4]$
cannot contain a global minimizer and can be safely discarded.

Interval Constraint Programming (ICP) aims at solving systems of
nonlinear equations and numerical optimization problems. Stemming from Interval Analysis and Interval Constraint Programming communities, filtering/contr-
action algorithms [8] narrow the bounds of the variables without loss of solutions.

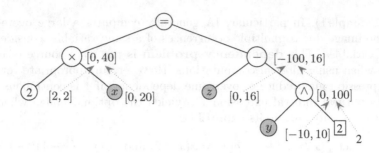

Fig. 1. HC4Revise: evaluation phase

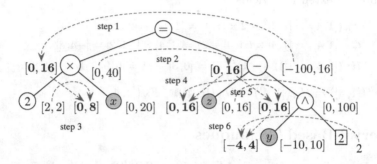

Fig. 2. HC4Revise: propagation phase

The standard contraction algorithm HC4Revise [9] carries out a double exploration of the syntax tree of a constraint to contract each occurrence of the variables (Example 3). It consists in an evaluation (bottom-up) phase that computes the elementary operation of each node, and a backward (top-down) propagation phase using inverse functions.

Example 3. Let $2x = z - y^2$ be an equality constraint, with $x \in [0, 20]$, $y \in [-10, 10]$ and $z \in [0, 16]$. The elementary expressions are the nodes $n_1 = 2x$, $n_2 = y^2$ and $n_3 = z - n_2$.

The evaluation phase (Fig. 1) computes $n_1 = 2 \times [0, 20] = [0, 40]$, $n_2 = [-10, 10]^2 = [0, 100]$ and $n_3 = [0, 16] - [0, 100] = [-100, 16]$.

The propagation phase (Fig. 2) starts by intersecting n_1 and n_3 (steps 1 and 2), then computes the inversion of each elementary expression (steps 3 to 6).

- steps 1 and 2: $n_1' = n_3' = n_1 \cap n_3 = [0, 40] \cap [-100, 16] = [0, 16]$
- step 3: $x' = x \cap \frac{n_1'}{2} = [0, 20] \cap [0, 8] = [0, 8]$
- step 4: $z' = z \cap (n_2 + n_3') = [0, 16] \cap ([0, 100] + [0, 16]) = [0, 16]$
- step 5: $n_2' = n_2 \cap (z' - n_3') = [0, 100] \cap ([0, 16] - [0, 16]) = [0, 16]$
- step 6: $y' = \square(y \cap -\sqrt{n_2'}, y \cap \sqrt{n_2'}) = \square([-4, 0], [0, 4]) = [-4, 4]$

The initial box $([0, 20], [-10, 10], [0, 16])$ has been reduced to $([0, 8], [-4, 4], [0, 16])$ without loss of solutions.

When partial derivatives are available, detecting **local monotonicity** with respect to a variable cancels the dependency effect due to this variable (Definition 3 and Example 4). In Definition 3, we call a monotonic variable a variable with respect to which f is monotonic.

Definition 3 (Monotonicity-based extension). *Let f be a function involving the set of variables $\mathcal{V}$. Let $\mathcal{X} \subseteq \mathcal{V}$ be a subset of k monotonic variables and $\mathcal{W} = \mathcal{V} \backslash \mathcal{X}$ the set of variables not detected monotonic. If x_i is an increasing (resp. decreasing) variable, we note $x_i^- = \underline{x_i}$ and $x_i^+ = \overline{x_i}$ (resp. $x_i^- = \overline{x_i}$ and $x_i^+ = \underline{x_i}$). f_{min} and f_{max} are functions defined by:*

$$f_{min}(\mathcal{W}) = f(x_1^-, \ldots, x_k^-, \mathcal{W})$$
$$f_{max}(\mathcal{W}) = f(x_1^+, \ldots, x_k^+, \mathcal{W})$$

The monotonicity-based extension F_M of f computes:

$$F_M = [\underline{f_{min}(\mathcal{W})}, \overline{f_{max}(\mathcal{W})}]$$

Example 4. Let $f(x) = x^2 - 2x$ and $X = [1,4]$. As seen in Example 1, $F_N([1,4]) = [-7,14]$. The derivative of f is $f'(x) = 2x - 2$, and $F'_N([1,4]) = 2 \times [1,4] - 2 = [0,6] \geq 0$. f is thus increasing with respect to x in X. Therefore, the monotonicity-based interval extension computes the optimal range: $F_M([1,4]) = [\underline{F(X)}, \overline{F(\overline{X})}] = [\underline{F(1)}, \overline{F(4)}] = [-1,8] = f([1,4])$.

This powerful property has been exploited in the contractor Mohc [10] and implemented in Charibde to enhance constraint propagation. However, the efficiency of this approach remains limited because the computation of partial derivatives may also be subject to overestimation.

4   Charibde Algorithm

We consider the following n-dimensional optimization problem and we assume that f is differentiable and that the analytical forms of f and its partial derivatives are available. We note n the dimension of the search-space.

$$\min_{x \in D \subset \mathbb{R}^n} f(x)$$
$$\text{s.t. } g_i(x) \leq 0, \ i \in \{1, \ldots, m\}$$

The current work extends the core method described in [6], in which we combined a GA and an IB&B that ran independently, and cooperated by exchanging information through shared memory in order to accelerate the convergence. In this framework, the GA quickly finds satisfactory solutions that improve the upper bound $\tilde{f}$ of the global minimum, and allows the IB&B to prune parts of the search-space more efficiently.

The interval-based algorithm embedded in Charibde follows a **Branch & Contract** (IB&C) scheme (described in Algorithm 1), namely an IB&B algorithm that integrates a contraction step based on HC4Revise. An IB&B merely

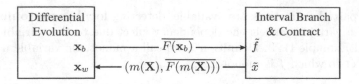

Fig. 3. Cooperative scheme of Charibde

relies on the refutation principle (discard a box if it is unfeasible or if it cannot contain a global minimizer). An IB&C may contract boxes by taking into account the constraints $g_i(x) \leq 0, i \in \{1, \ldots, m\}$ (feasibility) or $\frac{\partial \tilde{f}}{\partial x_i} = 0, i \in \{1, \ldots, n\}$ (local optimality) and $f \leq \tilde{f}$. Exploiting the analytical form of the objective function and its derivatives achieves faster convergence of the hybrid algorithm. Filtering algorithms show particular efficiency when $\tilde{f}$ is a good approximation of the global minimum provided by the EA thread, hence the necessity to quickly find an incumbent solution. Charibde thus outperforms our previous algorithm by far.

We note $\tilde{x}$ the best known solution, such that $\overline{F(\tilde{x})} = \tilde{f}$. The cooperation between the two threads boils down to three main steps (Fig. 3):

- Whenever the best known DE evaluation is improved, the best individual $\mathbf{x}_b$ is evaluated using IA. The upper bound of the image $\overline{F(\mathbf{x}_b)}$ – an upper bound of the global minimum – is sent to the IB&C thread
- In the IB&C algorithm, $\overline{F(\mathbf{x}_b)}$ is compared to the current best upper bound $\tilde{f}$. An improvement of the latter leads to a more efficiently pruning of the subspaces that cannot contain a (feasible) global minimizer
- Whenever the evaluation of the center $m(\mathbf{X})$ of a box improves $\tilde{f}$, the individual $m(\mathbf{X})$ replaces the worst individual $\mathbf{x}_w$ of DE, thus preventing premature convergence

In the following, we detail the implementations of the two main components of our algorithm: the deterministic IB&C thread and the stochastic DE thread.

4.1   Interval Branch & Contract Thread

We note $\mathcal{L}$ the priority queue in which the remaining boxes are stored and ε the desired precision. The basic framework of IB&C algorithms is described in Algorithm 1. The following rules have been experimentally tested and implemented in Charibde:

Selection rule: The box $\mathbf{X}$ for which $\overline{F(\mathbf{X})}$ is the largest is extracted from $\mathcal{L}$
Bounding rule: Evaluating $F(\mathbf{X})$ yields a rigorous enclosure of $f(\mathbf{X})$
Cut-off test: If $\tilde{f} - \varepsilon < \underline{F(\mathbf{X})}$, $\mathbf{X}$ is discarded as it cannot improve $\tilde{f}$ by more than ε
Midpoint test: If the evaluation of the midpoint of $\mathbf{X}$ improves $\tilde{f}$, $\tilde{f}$ is updated

Branching rule: X is bisected along the k-th dimension, where k is chosen according to the round-robin method (one dimension after another). The two resulting subboxes are inserted in $\mathcal{L}$ to be processed at a later stage

---

Algorithm 1. Interval Branch and Contract framework

$\tilde{f} \leftarrow +\infty$ ▷ best found upper bound
$\mathcal{L} \leftarrow \{\mathbf{X}_0\}$ ▷ priority queue of boxes to process
repeat
    Extract a box **X** from $\mathcal{L}$ ▷ selection rule
    Compute $F(\mathbf{X})$ ▷ bounding rule
    if X cannot be eliminated **then**   ▷ cut-off test
        Contract$(\mathbf{X}, \tilde{f})$ ▷ filtering algorithms
        Compute $F(m(\mathbf{X}))$ to update $\tilde{f}$ ▷ midpoint test
        Bisect **X** into $\mathbf{X}_1$ and $\mathbf{X}_2$ ▷ branching rule
        Store $\mathbf{X}_1$ and $\mathbf{X}_2$ in $\mathcal{L}$
    end if
until $\mathcal{L} = \varnothing$
return $(\tilde{f}, \tilde{x})$

4.2 Differential Evolution Thread

Differential Evolution (DE) is an EA that combines the coordinates of existing individuals with a particular probability to generate new potential solutions [11]. It has shown great potential for solving difficult optimization problems, and has few control parameters. Let us denote NP the population size, $W > 0$ the weighting factor and $CR \in [0, 1]$ the crossover rate. For each individual **x** of the population, three other individuals **u**, **v** and **w**, all different and different from **x**, are randomly picked in the population. The newly generated individual $\mathbf{y} = (y_1, \ldots, y_j, \ldots, y_n)$ is computed as follows:

$$y_j = \begin{cases} u_j + W \times (v_j - w_j) & \text{if } j = R \text{ or } rand(0,1) < CR \\ x_j & \text{otherwise} \end{cases} \tag{1}$$

R is a random index in $\{1, \ldots, n\}$ ensuring that at least one component of **y** differs from that of **x**. **y** replaces **x** in the population if $f(\mathbf{y}) < f(\mathbf{x})$.

Boundary constraints: When a component y_j lies outside the bounds $[\underline{D_j}, \overline{D_j}]$ of the search-space, the *bounce-back method* [12] replaces y_j with a component that lies between u_j (the j-th component of **u**) and the admissible bound:

$$y_j = \begin{cases} u_j + rand(0,1)(\overline{D_j} - u_j), & \text{if } y_j > \overline{D_j} \\ u_j + rand(0,1)(\underline{D_j} - u_j), & \text{if } y_j < \underline{D_j} \end{cases} \tag{2}$$

Evaluation: Given inequality constraints $\{g_i \mid i = 1, \ldots, m\}$, the evaluation of an individual **x** is computed as a triplet $(f_{\mathbf{x}}, n_{\mathbf{x}}, s_{\mathbf{x}})$, where $f_{\mathbf{x}}$ is the objective value, $n_{\mathbf{x}}$ the number of violated constraints and $s_{\mathbf{x}} = \sum_{i=1}^{m} \max(g_i(\mathbf{x}), 0)$. If at least one of the constraints is violated, the objective value is not computed

Selection: Given the evaluation triplets $(f_\mathbf{x}, n_\mathbf{x}, s_\mathbf{x})$ and $(f_\mathbf{y}, n_\mathbf{y}, s_\mathbf{y})$ of two candidate solutions $\mathbf{x}$ and $\mathbf{y}$, the best individual to be kept for the next generation is computed as follows:
  - if $n_\mathbf{x} < n_\mathbf{y}$ or $(n_\mathbf{x} = n_\mathbf{y} > 0$ and $s_\mathbf{x} < s_\mathbf{y})$ or $(n_\mathbf{x} = n_\mathbf{y} = 0$ and $f_\mathbf{x} < f_\mathbf{y})$ then $\mathbf{x}$ is kept
  - otherwise, $\mathbf{y}$ replaces $\mathbf{x}$

Termination: The DE has no termination criterion and stops only when the IB&C thread has reached convergence

5   Experimental Results

Charibde has been tested on a benchmark of standard test functions including quadratic, polynomial and nonlinear functions: bound-constrained problems (Rana, Egg Holder, Schwefel, Rosenbrock, Rastrigin, Michalewicz and Griewanlk) and inequality-constrained problems (Tension, Himmelblau, Welded Beam and Keane). Both the best known minimum in the literature and the certified global minimum[1] computed by Charibde are reported in Table 1. The global minima may be analytically computed for some separable or trivial functions, but for others (Rana and Egg Holder functions) no result concerning deterministic methods exists in the literature. Charibde has achieved new optimality results for three functions (Rana, Egg Holder and Michalewicz) and has proven the optimality of the known minima of the other functions.

Table 1. Test functions with best known and certified minima

	n	Type	Reference	Best known minimum	Certified minimum by Charibde
Bound-Constrained Problems					
Rana	4	Nonlinear	[15]	–	**−1535.1243381**
Egg Holder	10	Nonlinear	[16]	−8247 [17]	**−8291.2400675249**
Schwefel	10	Nonlinear		−4189.828873 [18]	−4189.8288727
Rosenbrock	50	Quadratic		0	0
Rastrigin	50	Nonlinear		0	0
Michalewicz	75	Nonlinear		–	**−74.6218111876**
Griewank	200	Nonlinear		0	0
Inequality-Constrained Problems					
Tension	3	Polynomial	[19]	0.012665232788319 [20]	0.0126652328
Himmelblau	5	Quadratic	[19]	−31025.560242 [21]	−31025.5602424972
Welded Beam	4	Nonlinear	[19]	1.724852309 [22]	1.7248523085974
Keane	5	Nonlinear	[23]	−0.634448687 [24]	−0.6344486869

Note that the constraints of Keane's function do not contain variables with multiple occurrences, and are therefore not subject to dependency. However, the first inequality constraint, describing a hyperbola in two dimensions, is active at

---

[1] Corresponding solutions are available upon request.

the global minimizer. The second inequality constraint is linear and is not active at the global minimizer. These constraints are highly combinatorial due to the sum and product operations, which makes constraint propagation rather ineffi-cient. The Egg Holder (resp. Rana) function is strongly subject to dependency: x_1 and x_n occur three (resp. Five) times in its expression, and $(x_2, \ldots, x_{n-1})$ occur six (resp. Ten) times. Their natural interval extensions therefore produce a large overestimation of the actual range. They are extremely difficult for interval-based solvers to optimize.

Partial derivatives of the objective function are computed using automatic differentiation [13]. To compute the partial derivatives of the functions that contain absolute values (Rana, Egg Holder, Schwefel and Keane), we use an interval extension based on the subderivative of $|\cdot|$ [14]:

$$|\cdot|'(X) = \begin{cases} [-1,-1] & \text{if } \overline{X} < 0 \\ [1,1] & \text{if } \underline{X} > 0 \\ [-1,1] & \text{otherwise} \end{cases} \tag{3}$$

The statistics of Charibde over 100 runs are presented in Table 2. ε is the numerical precision of the certified minimum such that $\tilde{f} - f^* < \varepsilon$, (NP, W, CR) are the DE parameters, t_{max} is the maximal computation time (in seconds), S_{max} is the maximal size of the priority queue $\mathcal{L}$, ne_f is the number of evaluations of the real-valued function f and $ne_F = ne_F^{DE} + ne_F^{IB\&C}$ is the number of evaluations of the interval function F computed in the DE thread (ne_F^{DE}) and the IB&C thread $(ne_F^{IB\&C})$. Note that ne_F^{DE} represents the number of improvements of the best DE evaluation. Because the DE thread keeps running as long as the IB&C thread has not achieved convergence, ne_f is generally much larger than

Table 2. Average results over 100 runs

	n	ε	NP	W	CR	t_{max}	S_{max}	ne_f	ne_F
Bound-Constrained Problems									
Rana	4	10^{-6}	50	0.7	0.5	222	42	274847000	$47 + 27771415$
Egg Holder	10	10^{-6}	50	0.7	0.5	768	45	423230200	$190 + 423230200$
Schwefel	10	10^{-6}	40	0.7	0.5	2.3	32	1462900	$150 + 362290$
Rosenbrock	50	10^{-12}	40	0.7	0.9	3.3	531	368028	$678 + 664914$
Rastrigin	50	10^{-15}	40	0.7	0	0.3	93	29372	$29 + 42879$
Michalewicz	75	10^{-9}	70	0.5	0	138	187	6053495	$1203 + 5796189$
Griewank	200	10^{-12}	50	0.5	0	11.8	134	188340	$316 + 116624$
Inequality-Constrained Problems									
Tension	3	10^{-9}	50	0.7	0.9	3.8	80	1324026	$113 + 1057964$
Himmelblau	5	10^{-9}	50	0.7	0.9	0.07	139	12147	$104 + 36669$
Beam	4	10^{-12}	50	0.7	0.9	2.2	11	316966	$166 + 54426$
Keane	5	10^{-4}	40	0.7	0.5	472	23	152402815	$125 + 99273548$

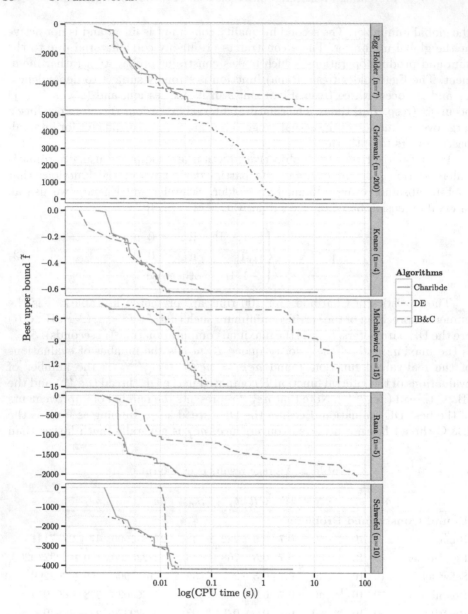

Fig. 4. Comparison of Charibde and standalone DE and IB&C (logarithmic x scale)

the number of evaluations required to reach $\tilde{f}$. These statistics suggest that the Egg Holder function, Keane's function and Rana's function are among the most challenging nonlinear problems for numerical solvers.

Figure 4 portrays the average comparison of performance between Charibde and standalone DE and IB&C over 100 runs of Six of the test functions (Egg Holder, Griewank, Keane, Michalewicz, Rana and Schwefel). A particular instance of each problem has been selected so that the standalone IB&C reaches convergence within reasonable time; to this end, the standard "best-first search" heuristic (extract the box $\mathbf{X}$ with the lowest $F(\mathbf{X})$) seemed more suitable. The DE algorithm reaches the global minimum for all instances. The IB&C generally experiences several phases of stagnation: this is due to the (crude) upper bounds of f^* obtained when evaluating the center of the boxes. On the contrary, Charibde benefits from the start of convergence of either the DE algorithm (Egg Holder, Keane, Rana and Schwefel) or the IB&C algorithm (Griewank and Michalewicz) to reach the global minimum faster than its standalone methods. Charibde proves to be highly competitive with the IB&C algorithm: on these (relatively simple) instances, the gain ratios in CPU time are respectively 2.04 (Egg Holder), 3.93 (Griewank), 1.14 (Keane), 377 (Michalewicz) and 3.95 (Rana). The IB&C algorithm however turns out to be more efficient than Charibde on the Schwefel function (gain ratio in CPU time: 0.36).

6 Conclusion

Extending the basic concept of [6], we have presented in this paper a new cooperative hybrid algorithm, Charibde, in which a stochastic Differential Evolution algorithm (DE) cooperates with a deterministic Interval Branch and Contract algorithm (IB&C). The DE algorithm quickly finds incumbent solutions that help the IB&C to improve pruning the search-space using interval propagation techniques. Whenever the IB&C improves the best known upper bound $\tilde{f}$ of the global minimum f^*, the corresponding solution is used as a new DE individual to avoid premature convergence toward local optima.

We have demonstrated the efficiency of this algorithm on a benchmark of difficult multimodal functions. Previously unknown results have been presented for Rana, Egg Holder and Michalewicz functions, while other known minima have been certified. By preventing premature convergence in the DE algorithm and providing the IB&C algorithm with a good approximation $\tilde{f}$ of f^*, Charibde proves highly competitive with its two standalone components.

References

1. Moore, R.E.: Interval Analysis. Prentice-Hall, Englewood Cliffs (1966)
2. Sotiropoulos, G.D., Stavropoulos, C.E., Vrahatis, N.M.: A new hybrid genetic algorithm for global optimization. In: Proceedings of Second World Congress on Nonlinear Analysts, pp. 4529–4538. Elsevier Science Publishers Ltd. (1997)
3. Holland, J.H.: Adaptation in Natural and Artificial Systems. University of Michigan Press, Ann Arbor (1975)
4. Zhang, X., Liu, S.: A new interval-genetic algorithm. Int. Conf. Nat. Comput. 4, 193–197 (2007)
5. Lei, Y., Chen, S.: A reliable parallel interval global optimization algorithm based on mind evolutionary computation. In: 2012 Seventh ChinaGrid Annual Conference, pp. 205–209 (2009)
6. Alliot, J.M., Durand, N., Gianazza, D., Gotteland, J.B.: Finding and proving the optimum: cooperative stochastic and deterministic search. In: 20th European Conference on Artificial Intelligence (2012)

7. Hansen, E.: Global Optimization Using Interval Analysis. Dekker, New York (1992)
8. Chabert, G., Jaulin, L.: Contractor programming. Artif. Intell. **173**, 1079–1100 (2009)
9. Benhamou, F., Goualard, F., Granvilliers, L., Puget, J.F.: Revising hull and box consistency. In: International Conference on Logic Programming, pp. 230–244. MIT press, Cambridge (1999)
10. Araya, I., Trombettoni, G., Neveu, B.: Exploiting monotonicity in interval constraint propagation. In: Proceedings of the AAAI, pp. 9–14 (2010)
11. Storn, R., Price, K.: Differential evolution - a simple and efficient heuristic for global optimization over continuous spaces. J. Global Optim. **11**, 341–359 (1997)
12. Price, K., Storn, R., Lampinen, J.: Differential Evolution - A Practical Approach to Global Optimization. Natural Computing. Springer, New York (2006)
13. Rall, L.B.: Automatic Differentiation: Techniques and Applications. LNCS, vol. 120. Springer, Heidelberg (1981)
14. Kearfott, R.B.: Interval extensions of non-smooth functions for global optimization and nonlinear systems solvers. Computing **57**, 57–149 (1996)
15. Whitley, D., Mathias, K., Rana, S., Dzubera, J.: Evaluating evolutionary algorithms. Artif. Intell. **85**, 245–276 (1996)
16. Mishra, S.K.: Some new test functions for global optimization and performance of repulsive particle swarm method. Technical report, University Library of Munich, Germany (2006)
17. Sekaj, I.: Robust parallel genetic algorithms with re-initialisation. In: Yao, X., et al. (eds.) PPSN 2004. LNCS, vol. 3242, pp. 411–419. Springer, Heidelberg (2004)
18. Kim, Y.H., Lee, K.H., Yoon, Y.: Visualizing the search process of particle swarm optimization. In: Proceedings of the 11th Annual Conference on Genetic and Evolutionary Computation, pp. 49–56. ACM (2009)
19. Coello Coello, C.A.: Use of a self-adaptive penalty approach for engineering optimization problems. Comput. Ind. **41**, 113–127 (1999)
20. Zhang, J., Zhou, Y., Deng, H.: Hybridizing particle swarm optimization with differential evolution based on feasibility rules. In: ICGIP 2012, vol. 8768 (2013)
21. Aguirre, A., Muñoz Zavala, A., Villa Diharce, E., Botello Rionda, S.: COPSO: constrained optimization via PSO algorithm. Technical report, CIMAT (2007)
22. Duenez-Guzman, E., Aguirre, A.: The Baldwin effect as an optimization strategy. Technical report, CIMAT (2007)
23. Keane, A.J.: A brief comparison of some evolutionary optimization methods. In: Proceedings of the Conference on Applied Decision Technologies. Modern Heuristic Search Methods, Uxbridge, 1995, pp. 255–272. Wiley, Chichester (1996)
24. Mishra, S.K.: Minimization of Keane's bump function by the repulsive particle swarm and the differential evolution methods. Technical report, North-Eastern Hill University, Shillong (India) (2007)

Local Optima Networks of the Permutation Flow-Shop Problem

Fabio Daolio[1]([⊠]), Sébastien Verel[2], Gabriela Ochoa[3], and Marco Tomassini[1]

[1] Department of Information Systems, University of Lausanne,
Lausanne, Switzerland
{fabio.daolio,marco.tomassini}@unil.ch
[2] Université du Littoral Côte D'Opale, LISIC, Dunkirk, France
verel@lisic.univ-littoral.fr
[3] Computing Science and Mathematics, University of Stirling, Scotland, UK
gabriela.ochoa@cs.stir.ac.uk

Abstract. This article extracts and analyzes local optima networks for the permutation flow-shop problem. Two widely used move operators for permutation representations, namely, swap and insertion, are incorporated into the network landscape model. The performance of a heuristic search algorithm on this problem is also analyzed. In particular, we study the correlation between local optima network features and the performance of an iterated local search heuristic. Our analysis reveals that network features can explain and predict problem difficulty. The evidence confirms the superiority of the insertion operator for this problem.

1 Introduction

The number and distribution of local optima in a combinatorial search space are known to impact the search difficulty on the corresponding landscape. Understating these features can also inform the design of efficient search algorithms. For example, it has been observed in many combinatorial landscapes that local optima are not randomly distributed, rather they tend to be relatively close to each other (in terms of a plausible metric) and to the known global optimum; clustered in a "central massif" (or "big valley" if we are minimizing) [4,11,18]. Search algorithms exploiting this globally convex structure have been proposed [4,18].

A recently proposed model of combinatorial fitness landscape *local optima networks*, captures in detail the distribution and topology of local optima in a landscape. The model was adapted from the study of energy landscapes in physics, which exist in continuous space [21]. In this network view of energy surfaces, vertices are energy minima and there is an edge between two minima if the system can jump from one to the other with an energy cost of the order of the thermal energies. In the combinatorial counterpart, vertices correspond to solutions that are minima or maxima of the associated combinatorial problem, but edges are defined differently, and are oriented and weighted. In a first version,

© Springer International Publishing Switzerland 2014
P. Legrand et al. (Eds.): EA 2013, LNCS 8752, pp. 41–52, 2014.
DOI: 10.1007/978-3-319-11683-9_4

the weights represent an approximation to the probability of transition between the respective basins in a given direction [6,16,23,25]. This definition, although informative, produced densely connected networks and required exhaustive sampling of the basins of attraction. A second version, *escape edges* was proposed in [24], which does not require a full computation of the basins. Instead, these edges account for the chances of escaping a local optimum after a controlled mutation (e.g.1 or 2 bit-flips in binary space) followed by hill-climbing. This second type of edges has, up to now, only been explored for binary spaces [24]. Also, previous work on networks with both basin and escape edges considered a single move operator on the corresponding search space.

This article extracts, analyzes and compares local optima networks of the *Permutation Flow-shop Problem* considering two types of move operators commonly used for permutation representation, namely, *insertion* and *exchange*. The article goes further and studies correlations among network features and the performance of an iterated local search heuristic.

2 Methods

2.1 Permutation Flow-Shop Problem

This section describes the optimization problem, solution representation, and move operators considered in this study.

Problem Formulation. In the *Permutation Flow-shop Problem* (PFSP), a flow of n jobs has to be scheduled for processing on m different machines in sequential order. Each of the n jobs will start at machine 1 and end at machine m. Concurrency and preemption are not allowed. In other words, job i can not start on machine $j + 1$ until machine j has completed it, and execution must run to completion once started. For any operation, job i will require a given processing time d_{ij} on machine j. Hence, a solution to the PFSP is a job processing order π, i.e. a permutation of the sequence of n jobs, where $\pi(i)$ denotes the ith job in the sequence. The objective is to find the permutation π_{best} yielding the minimum *makespan*, C_{max}, which is defined as the earliest completion time of its last job, $\pi_{best}(n)$, on the last machine m.

Search Operators. Several methods for solving the PFSP have been proposed [19], many of which are based on local search heuristics. For those, the choice of a move operator determines the topology of the search space [10]. We consider here two widely used operators for permutation representation. Namely, the *swap* (or *exchange*) operator, and the *shift* (or *insertion*) operator. $Exchange(x, y)$ simply swaps the job at positions x and y, while $Insert(x, y)$ selects a job at position x and inserts it into position y, shifting all others jobs; this operator is known to work well on the PFSP [22].

2.2   Local Optima Networks

This section overviews relevant definitions for building Local Optima Networks with Escape Edges in the presence of a neutral fitness landscapes.

A **fitness landscape** [20] is a triplet (S, V, f) where S, a search space, is the set of all admissible solutions, $V : S \longrightarrow 2^{|S|}$, a neighborhood structure, is the function that assigns to every $s \in S$ a set of neighbors $V(s)$, and $f : S \longrightarrow \mathbb{R}$ is a fitness function that maps the quality of the corresponding solutions.

Given a fitness landscape (S, V, f), a **local optimum** (LO), which is taken to be a maximum here, is a solution s^* such that $\forall s \in V(s)$, $f(s) \leq f(s^*)$.

In our study, the search space is composed of job sequences π of length n, therefore $|S| = n!$. The neighborhood is defined by the two selected move operators, consequently $|V(\pi)| = n(n-1)/2$ under the *exchange* operator and $|V(\pi)| = (n-1)(n-1)$ under the *insertion* operator. Finally, $f(\pi) = -C_{max}(\pi)$ that is to be maximized.

A **neutral neighbor** of s is a configuration $x \in V(s)$ with the same fitness value $f(x) = f(s)$; the size of the set $V_n(s) = \{x \in V(s) \mid f(x) = f(s)\}$ gives the **neutral degree** of a solution, i.e. how many neutral neighbors it has. When this number is high, the landscape can be thought of as composed of several sub-graphs of configurations with the same fitness value. This is the case for the fitness landscape of PFSP [14].

A neutral network (connected sub-graph whose vertices are neutral neighbors), also called a plateau, is a **local optimum neutral network** if all of its vertices are local optima.

Algorithm 1. Stochastic Best-Improvement Hill-Climber

Choose initial solution $s \in S$;
repeat
  randomly choose s' from $\{z \in V(s) | f(z) = \max_{x \in V(s)} f(x)\}$;
  if $f(s') \geq f(s)$ **then**
    $\lfloor \; s \leftarrow s'$;
until s *is in a Local Optimum Neutral Network*;

Since the size of the landscape is finite, we can mark the local optima neutral networks as $LONN_1, LONN_2, \ldots, LONN_n$. These are the vertices of the *local optima network* in the neutral case. In other words, we have a network whose nodes are themselves networks.

Algorithm 1 finds the local optima and defines their basins of attraction [16]. The connections among optima represent the chances of escaping from a $LONN$ and jumping into another basin after a controlled move [24]. But in a neutral landscape, the partition of solutions into basins of attraction is not sharp: Algorithm 1 is a stochastic operator h and $\forall s \in S$ there is a probability $p_i(s) = P(h(s) \in LONN_i)$. Therefore, the **basin of attraction** of $LONN_i$ is the set $b_i = \{s \in S \mid p_i(s) > 0\}$ and its size is $\sum_{s \in S} p_i(s)$ [25]. If we perturb a solution $s \in LONN_i$ by applying D random moves, we obtain a solution s' that will belong to another basin b_j with probability p_j, i.e. with probability p_j, $h(s')$ will eventually climb to $LONN_j$. The probability to go from s to b_j is then $p(s \rightarrow b_j) = \sum_{s' \in b_j} p(s \rightarrow s') p_j(s')$, where $p(s \rightarrow s') = P(s' \in \{z \mid d(z,s) \leq D\})$ is the probability for s' to be within D moves from s and can be evaluated in terms of relative frequency. Escaping from $LONN_i$ to the basin b_j after such a perturbation thus happens with probability $w_{ij} = p(LONN_i \rightarrow b_j) = \frac{1}{\sharp LONN_i} \sum_{s \in LONN_i} p_i(s) p(s \rightarrow b_j)$. Notice that w_{ij} might be different from w_{ji}.

The **Local Optima Network** (LON) is the weighted graph $G = (N, E)$ where the nodes are the local optima neutral networks, and there is an arc $e_{ij} \in E$ with weight $w_{ij} = p(LONN_i \rightarrow b_j)$ between two nodes i and j if $p(LONN_i \rightarrow b_j) > 0$.

3   Local Optima Network Analysis

This section overviews the main topological features of the permutation flow-shop local optima networks. Networks were extracted for instances with $n = 10$ jobs and $m \in \{5, 6, 7, 8, 9, 10\}$ number of machines. Instances of the unstructured (random) class were generated using the problem generator proposed by Watson et al. [26], which is based on the well-known Taillard benchmark [22]. For each combination of n and m, 30 instances were generated and results are presented through box-and-whiskers plots, to illustrate the distribution of the different metrics.

Four LON models are considered, namely, combining two neighborhoods: exchange and insertion, with two values of edge-escape distances: $D = 1$ and $D = 2$. For building the models, local optima are obtained using Algorithm 1 with, respectively, exchange and insertion moves, whereas the escape-edges consider the exchange move for the 4 models. The Algorithms were implemented in C++ using the "ParadisEO" library [5]; data analysis and visualization use R [17] with the appropriate packages for network analysis and statistical computing.

Network size: Figure 1a shows that the number of local optima for all LON models increases with the number of machines. This is consistent with the observation that increasing the number of machines (number of constraints) makes the problem harder to solve. The number of optima does not depend on the edges model ($D = 1$, $D = 2$), therefore, the two subplots in Fig. 1a are exactly the same. Figure 1a also indicates that the *exchange* LON model has a larger number of nodes as compared with the *insertion* model, which confirms that insertion is a better operator for the PFSP.

Figure 1b shows the density of edges, defined as the ratio of the LON number of edges to such number in a complete graph. As expected, the LON models with $D = 2$ are more dense. The density decreases with the number of machines for all models, and it is higher for the insertion LONs.

Clustering coefficient: The *clustering coefficient* of a network is the average probability that two neighbors of a given node are also neighbors of each other. In the language of social networks, the friend of your friend is likely also to be

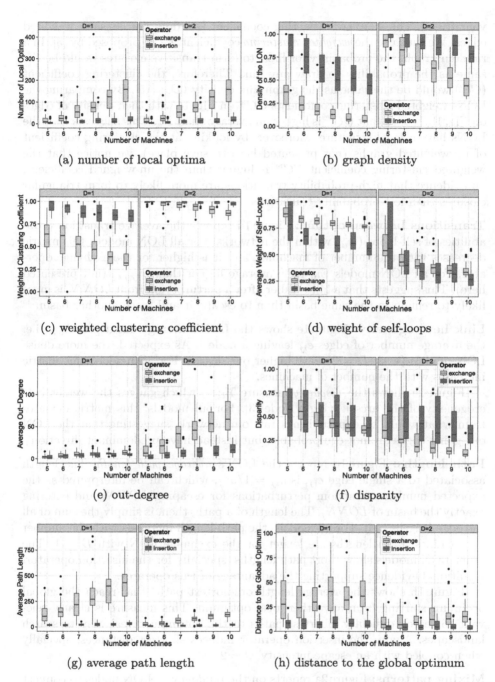

(a) number of local optima

(b) graph density

(c) weighted clustering coefficient

(d) weight of self-loops

(e) out-degree

(f) disparity

(g) average path length

(h) distance to the global optimum

Fig. 1. Box-and-whiskers plots giving the distribution of LON features. Boxes comprise the 0.25 and 0.75-quantiles, with a thick black line at the median value (i.e. the 0.50-quantile). Whiskers extend for 1.5 times the inter-quantile range and define "outliers" values, depicted as black dots.

your friend. The standard clustering coefficient [15] does not consider weighted edges. We thus use the *weighted clustering* coefficient, CC^w proposed by [3]. In a random graph, the probability for links to form transitive closures would be the same as the probability to draw any link. Therefore, the clustering coefficient (CC) would be the same as the graph link density (D_e) [15]. By comparing the LONs weighted clustering coefficients CC^w on Fig. 1c with their density of edges on Fig. 1b, we see that CC^w is higher on average than D_e. This suggests that the LONs have a local structure. Moreover, by looking at the clustering coefficient of un-weighted graphs (not presented here to save place), we notice that the weighted clustering coefficient CC^w is higher than the un-weighed coefficient, an evidence that high-probability transitions are more likely to form triangular closures than low-probability transitions.

Transitions between optima: Figure 1d reports the average transition probabilities of self-loops (w_{ii}) within the networks. For all LON models, this metric decreases with the number of machines, and it is higher for the exchange operator. For all LON models, w_{ii} is on average higher than $w_{ij,j\neq i}$ (not presented here). This suggests that a hill-climber after a perturbation from $LONN_i$ is more likely to remain in the same basin than to escape it and reach another basin.

Link heterogeneity: Figure 1e shows the LON's average out-degree k_{out}, i.e. the average number of edges e_{ij} leaving a node i. As expected, the more dense LON models (with $D = 2$) have higher out degree. For all models, this metric increases with the number of machines.

Figure 1e shows the *disparity* measure $Y_2(i)$, which gauges the weight heterogeneity of the arcs leaving a node [3]. For all models, this metric deviates from what would be expected in a random network, suggesting that the LON out-going edges are not equiprobable, but instead have predominant directions.

Path lengths: Figure 1g reports the LON's *average path length*. The length associated to a single edge e_{ij}, is $d_{ij} = 1/w_{ij}$, which can be interpreted as the expected number of random perturbations for escaping $LONN_i$ and entering exactly the basin of $LONN_j$. The length of a path, then, is simply the sum of all the edge lengths in it. For all models, the path length increases with the number of machines. Path lengths are longer for the exchange LON with $D = 1$. The other LON models show short path lengths, specially for the insertion operator. Additional evidence supporting the advantage of this operator.

Figure 1h shows the average length of shortest paths that reach the global optimum starting from any other local optimum. This measure is clearly relevant to search difficulty. Shortest paths to the optimum reveal easy to search landscapes. Again, the insertion operator induces shortest distances, specially when coupled with an escape intensity $D = 2$.

Mixing patterns: Figure 2a reports on the tendency of LON nodes to connect to nodes with similar degree. Specifically, figure shows the Newman's r coefficient, a common measure of *assortativity* roughly equivalent the Pearson correlation between the endpoints degree of all the existing links [15]. Degree mixing

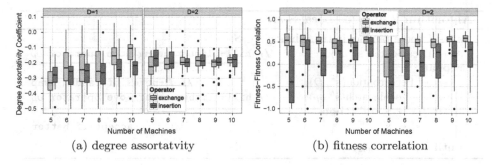

(a) degree assortatvity   (b) fitness correlation

Fig. 2. (a) Newman's r coefficient of assortativity; (b) Spearman ρ correlation between the fitness of a node and the weighted average of its neighbors' fitness.

is known to a have strong influence on the dynamical processes happening on complex network [2].

More interesting is to investigate mixing patterns with respect to the nodes fitness values. Figure 2b shows the Spearman correlation coefficient between the fitness of a $LONN_i$ and the average value of its neighbors $LONN_j$ fitnesses, weighted by the respective transition probabilities $w_{ij,j\neq i}$. This measure is less reliable on the small and dense LONs extracted from the *insertion* landscape, but on the *exchange* LONs, it suggests a positive fitness-fitness correlation that tends to increase with the number of machines. This might suggest that good solutions tend to be clustered within the search space.

More general and more pronounced is the positive correlation, measured by Spearman's ρ statistic, between the fitness value of a node and the sum of the weights of its incoming transitions. Considering all instances, ρ is in the 95 % confidence interval $(0.78, 0.81)$, indicating that the higher the fitness of a $LONN$, the easier it is to reach it. This is consistent with results on other combinatorial spaces displaying a positive correlation between fitness and basin size [7].

4   The Performance of Iterated Local Search

The network metrics studied in the previous section, suggest that the insertion operator is preferable over the exchange operator, and that an escape distance of 2 $(D = 2)$ induces an easier to search landscape. In order to corroborate these predictions, this section studies the performance of a heuristic search algorithm, specifically, iterated local search, when running on the modeled PFSP instances. Moreover, we show that it is possible to predict the running time of ILS using multi-linear regression model based on LON features.

Iterated local search is a relatively simple but powerful strategy, which operates by alternating between a perturbation stage and an improvement stage. This search principle has been rediscovered multiple times, within different research communities and with different names. The term *iterated local search* (ILS) was proposed in [12]. Algorithm 2 outlines the procedure.

Algorithm 2. Iterated Local Search

$s_0 \leftarrow$ Choose random initial solution $s \in S$;
$s^* \leftarrow$ LocalSearch(s_0, op);   `// hill-climber using move operator` op
repeat
  $s' \leftarrow$ Perturbation(s^*, D);   `// D-moves of random swap`
  $s'^* \leftarrow$ LocalSearch(s', op);   `// hill-climber using move operator` op
  if $f(s'^*) > f(s^*)$ **then**
    $s^* \leftarrow s'^*$;   `// accept if better`
until $FE \leq FE_{max}$;

The *LocalSearch* procedure in Algorithm 2, corresponds to the stochastic hill-climber given in Algorithm 1. In our implementation, the two operators studied: insertion and exchange can be used in this stage. The perturbation stage uses only the exchange operator but with two different intensities of one or two operator applications. Notice that Algorithm 2 follows closely the structure of basins of the search space, and thus, the LON models should explain the performance of such ILS. Specifically, four ILS implementations are tested, namely, using insertion and exchange in the local stage, and using one or two applications of the exchange operator in the perturbation stage, which we denote $D = 1$ and $D = 2$.

Experimental Setup: The same instances studied in Sect. 3 are considered, i.e. unstructured (random) instances with $n = 10$ jobs and $m \in \{5, 6, 7, 8, 9, 10\}$ number of machines. The four variants of ILS (Algorithm 2 described above) are tested. The maximum running time is set to $FE_{max} = 0.2|S| = 0.2 \cdot 10! = 725760$ function evaluations. On each instance, independent runs are randomly restarted 1000 times upon termination, which occurs either on finding the global optimum or on exhausting the FE budget.

For assessing the algorithms' performance, we use the expected number of function evaluations to reach the global optimum (*Run-Length* [9]), considering independent restarts of the ILS algorithms [1]. This accounts for both the success rate ($p_s \in (0, 1]$) and the convergence speed. After $(N - 1)$ unsuccessful runs stopped at T_{us}-steps and the final successful one running for T_s-steps, the total run-length would be $T = \sum_{k=1}^{N-1} (T_{us})_k + T_s$. Taking the expectation and considering that N follows a geometric distribution (Bernoulli trials) with parameter p_s, it gives: $\mathbb{E}(T) = \left(\frac{1-p_s}{p_s}\right) \mathbb{E}(T_{us}) + \mathbb{E}(T_s)$, where $\mathbb{E}(T_{us}) = FE_{max}$, the ratio of successful to total runs is an estimator for p_s, and $\mathbb{E}(T_s)$ can be estimated by the average running time of successful runs.

Comparing the Performance of ILS Variants: Figure 3 compares the performance of the four ILS variants. Figure 3a reports the estimated probability of success, which is clearly superior for ILS variants with perturbation strength of 2 ($D = 2$). In this case the ILS algorithm solves all instances to optimality in the median. For one perturbation ($D = 1$, in Fig. 3a), success rates are much lower, specially for the exchange operator, where they decrease with

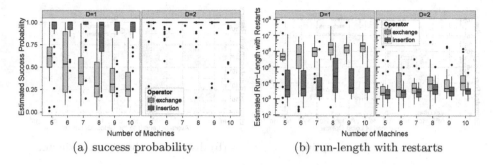

(a) success probability   (b) run-length with restarts

Fig. 3. Performance of iterated local search: (a) success probability, (b) run-length.

increasing number of machines. A closer look at the performance of the ILS variants is appreciated in Fig. 3b, showing the estimated run-lengths. Run-lengths are much higher for ILS variants with a single exchange $(D = 1)$. For both $D = 1$ and $D = 2$, the insertion operator produce shorter running lengths, although differences are greater when a single perturbation is used. Finally, for all ILS variants, the running length tends to increase with the number of machines. These performance observations, are consistent with the search difficulty predicted by the LON metrics in Sect. 3.

Table 1. Spearman's ρ statistic for the correlation between the estimated run-length of ILS variants and the LON metrics by the respective move and perturbation. N_v nb of local optima, CC^w avg weighted clustering coeff., F_{nn} neighboring nodes fitness-fitness corr., k_{nn} neighboring nodes degree-degree corr., r Newman's assortativity, L_{opt} avg shortest distance to the global optimum, L_v avg path length, Fs_{in} fitness-strength(in) corr., w_{ii} avg weight of self-loops, Y_2 avg disparity of (out)links, k_{out} avg (out)degree.

ILS/LON	N_v	CC^w	F_{nn}	k_{nn}	r	L_{opt}	L_v	Fs_ρ	w_{ii}	Y_2	k_{out}
Insertion $D1$	0.46	−0.221	0.199	0.078	0.238	0.634	0.40	−0.101	−0.31	−0.41	0.479
Insertion $D2$	0.54	−0.209	0.316	−0.165	0.117	0.691	0.45	−0.167	−0.476	−0.46	0.55
Exchange $D1$	0.535	−0.506	−0.004	0.142	0.353	0.624	0.536	−0.102	−0.235	−0.473	0.448
Exchange $D2$	0.408	−0.255	0.22	−0.111	0.165	0.527	0.353	−0.035	−0.272	−0.434	0.409

Performance prediction: This section explores the correlations between the LON metrics from Sect. 3 and the ILS performance presented above. More precisely, Table 1 reports the rank-based Spearman's ρ statistic between each LON metric and the ILS estimated run-length, considering the natural pairings of move operator and perturbation intensity between ILS variants and LON models. In all cases, the higher the number of local optima (N_v) and, even more importantly, the longer the average lengths of paths to the global optimum (L_{opt}), the longer it takes for the iterated search to solve an instance to optimality. Figure 4 shows such correlations, which are the highest observed. Other scatter plots are less clear and are left out for reasons of space, but admittedly, their interpretation would also be less straightforward.

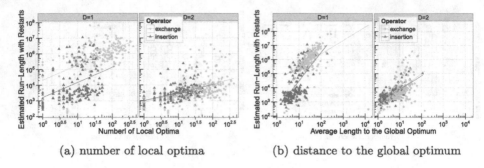

(a) number of local optima   (b) distance to the global optimum

Fig. 4. Scatter plots of the estimated run-length versus different network metrics.

Finally, in order to investigate how the LON features could be used to predict the search difficulty on the whole set of explored landscapes, we propose a set of linear regression models having the estimated run-length as a dependent variable, log-transformed after a preliminary analysis (log-likelihood of Box-Cox's power parameter). We perform an exhaustive search in the set of all possible regressors subsets [13] and for each subset size we retain the best model according to Mallow's C_p statistic [8]. Results are given in Table 2. Interestingly, the number of local optima N_v is never chosen; instead, the best single predictor is the average length of the shortest paths to the global optimum L_{opt}, log-transformed, which alone accounts for more than 57 % of the observed run-length variance across the PFSP instances under study.

Table 2. Exhaustive search among all regressors subsets for the multiple linear regression predicting the logarithm of estimated run-length as a function of the LON metrics. For each number of predictors $\sharp P$, the best model in terms of Mallow's C_p statistic is given, along with its estimated regression coefficients and the resulting adjusted R^2.

$\sharp P$	$\log(N_v)$	CC^w	F_{nn}	k_{nn}	r	$\log(L_{opt})$	$\log(L_v)$	Fs_ρ	w_{ii}	Y_2	k_{out}	C_p	$adj R^2$
1						2.13						265.54	0.574
2		−5.18				1.43						64.06	0.675
3						1.481	0.895				−0.042	16.48	0.700
4		−2.079				1.473	0.540				−0.032	8.75	0.704
5		−2.388			−1.633	1.470	0.528				−0.030	5.97	0.706
6		−2.532			−1.722	1.469	0.472	−1.405			−0.028	3.75	0.707
7		−2.772			−1.986	1.461	0.427	−1.497		−0.408	−0.029	5.02	0.707
8		−2.748	−0.188		−2.078	1.464	0.452	−1.579		−0.515	−0.029	6.39	0.707

5   Conclusions

This article extracts and analyzes, for the first time, the local optima networks of the permutation flow-shop problem. The LON model with the so-called escape-edges, which account for the chances of escaping a local optimum after a controlled perturbation (1 or 2 random exchanges in our implementation), is extended to landscapes with neutrality. Two move operators, widely used for permutation representations (exchange and insert), are considered and contrasted.

LONs induced by the insertion operator present fewer nodes (i.e. fewer local optima), and shortest distances both among nodes and from any node to the global optimum. This evidence supports the superior performance of the insertion over the exchange move as reported in the literature. The LON models with $D = 2$ produce shortest distances among nodes, and from any node to the global optimum, compared to models with $D = 1$. Therefore a local search heuristic using the insertion operator for adaptive walks and several kicks of the exchange operator to escape local optima, should perform well on these PFSP instances.

Indeed, four iterated local search variants were implemented and tested, which resemble the considered LON models. Among these, the ILS with insertion in the improvement stage and two exchanges in the perturbation stage, produced the best performance. This confirms the intuitions from the LON model metrics. Actually, not only the LON metrics correlate with the search performance, but also the ILS running time can be estimated using the LON features.

Future work will explore larger problems, which requires sampling to extract the LON models, and additional permutation flow-shop instance classes, such as machine-correlated and mixed-correlated instances [26]. The ultimate goal is to derive easy-to-compute landscape metrics that can predict the performance and guide the design of heuristic search algorithms when solving difficult combinatorial problems. This article is an additional step in this direction.

References

1. Auger, A., Hansen, N.: Performance evaluation of an advanced local search evolutionary algorithm. In: The 2005 IEEE Congress on Evolutionary Computation, 2005, vol. 2, pp. 1777–1784. IEEE (2005)
2. Barrat, A., Barthélemy, M., Vespignani, A.: Dynamical Processes on Complex Networks. Cambridge University Press, Cambridge (2008)
3. Barthélemy, M., Barrat, A., Pastor-Satorras, R., Vespignani, A.: Characterization and modeling of weighted networks. Phys. A Stat. Mech. Appl. **346**(1), 34–43 (2005)
4. Boese, K.D., Kahng, A.B., Muddu, S.: A new adaptive multi-start technique for combinatorial global optimizations. Oper. Res. Lett. **16**, 101–113 (1994)
5. Cahon, S., Melab, N., Talbi, E.G.: Paradiseo: A framework for the reusable design of parallel and distributed metaheuristics. J. Heuristics **10**, 357–380 (2004)
6. Daolio, F., Tomassini, M., Verel, S., Ochoa, G.: Communities of minima in local optima networks of combinatorial spaces. Phys. A Stat. Mech. Appl. **390**, 1684–1694 (2011)
7. Daolio, F., Verel, S., Ochoa, G., Tomassini, M.: Local optima networks of the quadratic assignment problem. In: 2010 IEEE Congress on Evolutionary Computation (CEC), pp. 1–8. IEEE (2010)
8. Gilmour, S.G.: The interpretation of mallows's C_p-statistic. The Statistician **45**, 49–56 (1996)
9. Hoos, H., Stützle, T.: Stochastic local search: Foundations and applications. Morgan Kaufmann, San Francisco (2005)

10. Jones, T.: Evolutionary algorithms, fitness landscapes and search. Ph.D. Thesis, The University of New Mexico (1995)
11. Kauffman, S., Levin, S.: Towards a general theory of adaptive walks on rugged landscapes. J. Theor. Biol. **128**, 11–45 (1987)
12. Lourenço, H.R., Martin, O., Stützle, T.: Iterated local search. In: Glover, F., Kochenberger, G. (eds.) Handbook of Metaheuristics, International Series in Operations Research and Management Science, vol. 57, pp. 321–353. Kluwer Academic Publishers, Norwell (2002)
13. Lumley, T., Miller, A.: Leaps: Regression subset selection (2009). http://CRAN. R-project.org/package=leaps
14. Marmion, M.-E., Dhaenens, C., Jourdan, L., Liefooghe, A., Verel, S.: On the neutrality of flowshop scheduling fitness landscapes. In: Coello, C.A.C. (ed.) LION 2011. LNCS, vol. 6683, pp. 238–252. Springer, Heidelberg (2011)
15. Newman, M.: The structure and function of complex networks. SIAM Rev. **45**, 167–256 (2003)
16. Ochoa, G., Verel, S., Tomassini, M.: First-improvement vs. best-improvement local optima networks of NK landscapes. In: Schaefer, R., Cotta, C., Kołodziej, J., Rudolph, G. (eds.) PPSN XI. LNCS, vol. 6238, pp. 104–113. Springer, Heidelberg (2010)
17. R Core Team: R: A Language and Environment for Statistical Computing. R Foundation for Statistical Computing, Vienna, Austria (2013). http://www.R-project. org/. ISBN 3-900051-07-0
18. Reeves, C.R.: Landscapes, operators and heuristic search. Ann. Oper. Res. **86**, 473–490 (1999)
19. Ruiz, R., Maroto, C.: A comprehensive review and evaluation of permutation flowshop heuristics. Eur. J. Oper. Res. **165**(2), 479–494 (2005)
20. Stadler, P.: Fitness landscapes. Biol. Evol. Stat. Phys. **585**, 183–204 (2002)
21. Stillinger, F.: A topographic view of supercooled liquids and glass formation. Science **267**, 1935–1939 (1995)
22. Taillard, E.: Some efficient heuristic methods for the flow shop sequencing problem. Eur. J. Oper. Res. **47**(1), 65–74 (1990)
23. Tomassini, M., Verel, S., Ochoa, G.: Complex-network analysis of combinatorial spaces: The NK landscape case. Phys. Rev. E **78**(6), 066114 (2008)
24. Verel, S., Daolio, F., Ochoa, G., Tomassini, M.: Local Optima Networks with Escape Edges. In: Procedings of International Conference on Artificial Evolution (EA-2011). pp. 10–23. Angers, France (Oct 2011).
25. Verel, S., Ochoa, G., Tomassini, M.: Local optima networks of NK landscapes with neutrality. IEEE Trans. Evol. Comput. **15**(6), 783–797 (2011)
26. Watson, J., Barbulescu, L., Whitley, L., Howe, A.: Contrasting structured and random permutation flow-shop scheduling problems: search-space topology and algorithm performance. INFORMS J. Comput. **14**(2), 98–123 (2002)

Linear Convergence of Evolution Strategies with Derandomized Sampling Beyond Quasi-Convex Functions

Jérémie Decock[✉] and Olivier Teytaud

TAO-INRIA, LRI, CNRS UMR 8623, Université Paris-Sud, Orsay, France
{jeremie.decock,olivier.teytaud}@inria.fr

Abstract. We study the linear convergence of a simple pattern search method on non quasi-convex functions on continuous domains. Assumptions include an assumption on the sampling performed by the evolutionary algorithm (supposed to cover efficiently the neighborhood of the current search point), the conditioning of the objective function (so that the probability of improvement is not too low at each time step, given a correct step size), and the unicity of the optimum.

1 Introduction

Continuous evolutionary algorithms are well known for robust convergence. However, most proven results are for simple objective functions, e.g. sphere functions [1]. Results also include compositions with monotone functions (so that not only convex functions are covered), but the considered objective functions are nonetheless still almost always quasi-convex (i.e. sublevel sets are convex), as well as most derivative free optimization algorithms [4], whereas nearly all testbeds are based on more difficult functions [7,11]. Extensions to non quasi-convex functions are still rare [12] and limited to convergence (i.e.: asymptotically we will find the optimum). We here extend such results to linear convergence (i.e. the precision after n iterations is $O(\exp(-\Omega(n)))$. There are works devoted to unimodal objective functions, without convexity assumptions [6], but such works are in the discrete domain and do not say anything for the linear convergence on continuous domains. All in all, only one of the six objective functions of Fig. 1 is covered by existing results, in terms of linear convergence.

In this paper, we prove linear convergence of a simple pattern search method with derandomized sampling on non quasi-convex families of functions. Section 2 presents the framework, and the assumptions under which our results hold. Section 3 is the mathematical analysis, under this set of assumptions. Section 4 presents the application to positive definite quadratic forms: it shows that the family of quadratic forms with conditioning bounded by some constant verifies our set of assumptions, and therefore that our evolution strategy with derandomized sampling has linear convergence rate on such objective functions. Incidentally, this section emphasizes the critical underlying assumptions for proving

© Springer International Publishing Switzerland 2014
P. Legrand et al. (Eds.): EA 2013, LNCS 8752, pp. 53–64, 2014.
DOI: 10.1007/978-3-319-11683-9_5

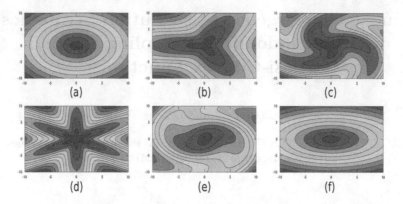

Fig. 1. Six graphical representations of easy objective functions; only the first one (sphere) is covered by existing linear convergence results. Even the sixth one (ellipsoids) is not included in published linear convergence results. We extend to all functions verifying Eqs. 1–6 (see these equations in text), including all functions presented here. We present assumptions under which our results hold in Sect. 2, the main result in Sect. 3, and we will show in details that the sixth case above (quadratic functions) is covered by the result in Sect. 4 (but case 1 is a special case of case 6, and cases 2, 3, 4, 5 can be tackled similarly). Reference [12] provides other examples, with different but very related assumptions; their examples are also covered by our theorem.

the result, suggesting extensions to other families of fitness functions. Section 5 concludes and discusses limitations and further work.

2 A Simple Pattern Search Method

We consider an evolutionary algorithm as in Algorithm 1. As the sampling is derandomized, we might indeed call this algorithm a pattern search method. We assume the followings.

The Objective Function. We assume that the function f has a unique minimum. Without loss of generality, we assume that the objective function verifies $f(0) = 0$ and that this is the minimum. The considered algorithms are invariant by transition or composition with monotone functions, so this does not reduce the generality of the analysis.

Conditioning. We assume that

$$K'||\boldsymbol{x}|| \le f(\boldsymbol{x}) \le K''||\boldsymbol{x}|| \tag{1}$$

for all $\boldsymbol{x}$ in $\mathbb{R}^d$ and for some constants $K' > 0$ and $K'' > 0$. We point out that, as we consider algorithms which are invariant under transformations of the objective function by composition with monotonic functions, this assumption is not so strong as a constraint and quadratic positive definite forms with bounded condition number are in fact covered (their square root verifies Eq. 1).

Algorithm 1. The Simple Evolution Strategy. In case there is no unicity for choosing x', any breaking tie solution is ok. (c) refers to the counting operation, which will be important in the proof. $[[1, k]]$ stands for the integer set $\{1, \ldots, k\}$.

Initialize $x \in \mathbb{R}^d$
Parameters $k \in \mathbb{N}^*, \delta_1, \ldots, \delta_k \in \mathbb{R}^d, \sigma \in \mathbb{R}_+^*, k_1 \in \mathbb{N}^*, k_2 \in \mathbb{N}^*$
for $t = 1, 2, 3, \ldots$ **do**

   //   just for archiving
   $X_t \leftarrow x$

   //   mutations
   For $i \in [[1, k]]$, $x_i \leftarrow x + \sigma \delta_i$

   //   useful auxiliary variables
   $n \leftarrow$ number of x_i such that $f(x_i) < f(x)$ (c)
   $x' \leftarrow x_i$ with $i \in [[1, k]]$ such that $f(x_i)$ is minimum

   //   step-size adaptation
   if $n \leq k_1$ **then**
     $\sigma \leftarrow \sigma/2$
   end if
   if $n \geq k_2$ **then**
     $\sigma \leftarrow 2\sigma$
   end if

   //   win: accepted mutation
   if $k_1 < n < k_2$ **then**
     $x \leftarrow x'$
   end if
end for

Good Sampling. Here we use the derandomized sampling assumptions (Eqs. 2–6), which are crucial in our work. This sampling is deterministic, as in pattern search methods [4]. We assume that for some $0 < b < b' \leq 2b' < c' \leq c, 0 < \eta < 1$ and $\forall x \in \mathbb{R}^d$,

$$\sigma \text{ too large: } \sigma \geq b^{-1}||x||$$
$$\Rightarrow \#\{i \in [[1, k]]; f(x + \sigma\delta_i) < f(x)\} \leq k_1 \qquad (2)$$

$$\sigma \text{ small enough: } \sigma \leq b'^{-1}||x||$$
$$\Rightarrow \#\{i \in [[1, k]]; f(x + \sigma\delta_i) < f(x)\} > k_1 \qquad (3)$$

$$\sigma \text{ large enough: } \sigma \geq c'^{-1}||x||$$
$$\Rightarrow \#\{i \in [[1, k]]; f(x + \sigma\delta_i) < f(x)\} < k_2 \qquad (4)$$

σ too small: $\sigma \leq c^{-1}||\boldsymbol{x}||$

$$\Rightarrow \#\{i \in [[1,k]]; f(\boldsymbol{x} + \sigma\boldsymbol{\delta}_i) < f(\boldsymbol{x})\} \geq k_2 \qquad (5)$$

Perfect σ : $b'^{-1}||\boldsymbol{x}|| \leq \sigma \leq c'^{-1}||\boldsymbol{x}||$

$$\Rightarrow \exists i \in [[1,k]]; f(\boldsymbol{x} + \sigma\boldsymbol{\delta}_i) \leq \eta f(\boldsymbol{x}) \qquad (6)$$

Discussion on Assumptions. Assumptions 2, 3, 4, 5, 6 basically assume that the sampling is regular enough for the shape of the level sets. For example, the finite VC-dimension of ellipsoids ensure that, when the conditioning is bounded, quadratic functions verify the assumptions above (and therefore the theorem below) with arbitrarily high probability if $\boldsymbol{\delta}_1,\dots,\boldsymbol{\delta}_k$ are randomly drawn and if k is large enough. Importantly, the critical assumption in the derandomization is that all iterations have the same $\boldsymbol{\delta}_1,\dots,\boldsymbol{\delta}_k$. This will be developed in Sect. 4.

Assumptions 1 and 6 use the fitness values; but they just have to hold for one of the fitness values obtained by replacing f with $g \circ f$ with g a monotone function.

3   Mathematical Analysis

Main Theorem: *Assume Eqs. 1–6. There exists a constant K, depending on $\eta, K', K'', \max_i ||\boldsymbol{\delta}_i||$ only such that for index t sufficiently large*

$$\ln(||\boldsymbol{X}_t||)/t \leq K < 0 \qquad (7)$$

(with $\ln(0) = -\infty$) where the sequence of $\boldsymbol{X}_t$ is defined as in Algorithm 1.

Proof: First, we briefly explain and illustrate the proof, before the formal proof below. The proof is sketched in Fig. 2. At each iteration t, we are at some point in the figure; the x-axis is $-\ln(||\boldsymbol{x}||)$ (equivalent to $-\ln(f(\boldsymbol{x}))$, by Eq. 1), the y-axis is $l = \ln\left(\frac{||\boldsymbol{x}||}{\sigma}\right)$. The step-size adaptation ensures that if we are at the bottom ($l \leq \ln(b)$), we go upwards; if we are at the top ($l \geq \ln(c)$), we go downwards. Between $l = \ln(b)$ and $l = \ln(c)$, everything can happen; but if there's no "win" case in the mean time, we will arrive between $l = \ln(b')$ and $l = \ln(c')$, where a win is ensured. As steps are fast, this can not take too much time (if there is no "win", l increases by $\ln(2)$ or decreases by $\ln(2)$ in direction of the "forced win" range $[\ln(b'), \ln(c')]$). This will be formalized below. $c' \geq 2b'$ ensures that the algorithm can not jump from $l < \ln(b')$ to $l > \ln(c')$ or from $l > \ln(c')$ to $l < \ln(b')$. Therefore there is necessarily a "win" in the mean time. Equation 6 ensures that wins provide a significant improvement.

We now write the proof formally. Consider an iteration of the algorithm, with n the number of mutations i with $f(\boldsymbol{x} + \sigma\boldsymbol{\delta}_i) < f(\boldsymbol{x})$ (as defined in Algorithm 1, Eq. (c)).

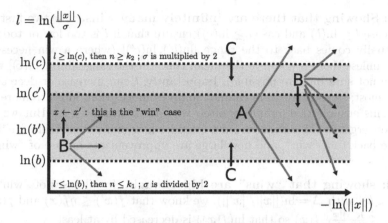

Fig. 2. The linear convergence proof in a nutshell. X-axis: $-\ln(\|x\|)$. Y-axis: $l = \ln(\frac{\|x\|}{\sigma})$. At each iteration, either **case A holds:** then the iteration is for sure an improvement by a factor at least η, or **case B holds:** the iteration can be an improvement or not; if not, the point is moved towards case A by $\ln(2)$ upwards or downwards, or **case C holds:** then x is moved upwards (if it is at the bottom) or downwards (if it is at the top). This ensures that after finitely many time steps we go back to case A unless there is a "win" by case B in the mean time. The crucial point for the proof is that each "win" is an improvement by least a controlled factor η, so that the slope of "win" arrows is bounded, so that there is linear convergence and not only an infinite sequence of "small" improvements.

Define $l = \ln\left(\frac{\|x\|}{\sigma}\right)$. Equations 2–6 can be rephrased as follows:

$$l \leq \ln(b) \Rightarrow \#\{i \in [[1,k]]; f(x + \sigma\delta_i) < f(x)\} \leq k_1 \qquad (8)$$

$$l \geq \ln(b') \Rightarrow \#\{i \in [[1,k]]; f(x + \sigma\delta_i) < f(x)\} > k_1 \qquad (9)$$

$$l \leq \ln(c') \Rightarrow \#\{i \in [[1,k]]; f(x + \sigma\delta_i) < f(x)\} < k_2 \qquad (10)$$

$$l \geq \ln(c) \Rightarrow \#\{i \in [[1,k]]; f(x + \sigma\delta_i) < f(x)\} \geq k_2 \qquad (11)$$

$$\ln(b') \leq l \leq \ln(c') \Rightarrow \exists i \in [[1,k]]; f(x + \sigma\delta_i) \leq \eta f(x) \qquad (12)$$

Define x' as in Algorithm 1. We get the following behavior:

- Forced increase: if $l \leq \ln(b)$, then $n \leq k_1$; σ is divided by 2, and l is increased by $\ln(2)$ (Eq. 8). This is a case C in Fig. 2.
- Forced decrease: if $l \geq \ln(c)$, then $n \geq k_2$; σ is multiplied by 2, and l is decreased by $\ln(2)$ (Eq. 11). This is a case C in Fig. 2.
  Forced win: if $\ln(b') \leq l \leq \ln(c')$, then $x \leftarrow x'$; this is the "sure win" case (Eq. 12); l can be increased (at most by $\max_i \|\delta_i\|$) or decreased (by $\Delta = \ln(\|x\|/\|x'\|)$). This is a case A in Fig. 2.

Importantly, these 3 cases do not cover all possible cases; $\ln(c') < l < \ln(c)$ and $\ln(b) < l < \ln(b')$ are not covered in items above. These two remaining cases are termed case B in Fig. 2.

Step 1: Showing that there are infinitely many wins. The two first lines above (case $l \leq \ln(b)$ and case $l \geq \ln(c)$) ensure that if l is too low or too high, it eventually comes back to the range $[\ln(b'), \ln(c')]$ (where a win necessarily occurs), unless there is a win in the mean time (in the range $[\ln(b), \ln(c)]$ where wins are not sure but are possible). Importantly, l can increase or decrease by $\ln(2)$ at most; so the algorithm can not jump from less than $\ln(b')$ to more than $\ln(c')$. This ensures that infinitely often we have a win $\boldsymbol{x} \leftarrow \boldsymbol{x}'$. But we want linear convergence. Therefore we must consider how many steps there are before we come back to a "win", and how large are improvements in case of "win".

Step 2: showing that "wins" are big enough. In all cases of "win", i.e. $k_1 < n < k_2$, with $\Delta = \ln(\|\boldsymbol{x}\|/\|\boldsymbol{x}'\|)$, we know that $f(\boldsymbol{x}') \leq \eta f(\boldsymbol{x})$ and $f(\boldsymbol{x}') \leq K''\|\boldsymbol{x}'\| \leq \frac{K''}{K'}\frac{\|\boldsymbol{x}'\|}{\|\boldsymbol{x}\|} f(\boldsymbol{x})$ so that $\ln(f(\boldsymbol{x}))$ is decreased by at least

$$\max(\ln(1/\eta), \ln(K'/K'') + \Delta). \tag{13}$$

After a "win", with $l' = \ln\left(\frac{\|\boldsymbol{x}'\|}{\sigma}\right)$,

– if $l' \leq \ln(b')$, then the number of iterations before the next win is at most $z = 1 + \ln(\frac{c}{b})\Delta/\ln(2)$, because $l' \geq \ln(b) - \Delta \geq \ln(b') - \ln(b'/b) - \Delta \geq \ln(b') - \ln(c/b) - \Delta$ and forced increase are by steps of at least $\ln(2)$.
– if $l' \geq \ln(c')$, then the number of iterations before the next win is at most $z = 1 + \ln(\frac{c}{b}) \max_i \ln(\|\boldsymbol{\delta}_i\|)/\ln(2)$, because $l' \leq \ln(c) - max_i \ln(\boldsymbol{\delta}_i) \leq \ln(c') - \ln(c'/c) - max_i \ln(\boldsymbol{\delta}_i) \leq \ln(c') - \ln(c/b) - \max_i \ln(\boldsymbol{\delta}_i)$ and forced decreases are by steps of at least $\ln(2)$.
– less than in both cases above, otherwise.

In both cases, Eq. 13 divided by z is lower bounded by some positive constant

$ProgressRate$
$=$Eq. 13 divided by z
$$= \frac{\max(\ln(1/\eta), \ln(K'/K'') + \Delta_i)}{\min(1 + \ln(c/b)\Delta_i/\ln(2), 1 + \ln(c/b)\max_j \ln(\|\boldsymbol{\delta}_j\|)/\ln(2))}. \tag{14}$$

Step 3: summing iterations. Equation 14 is the progress rate between two wins, after normalization by the number of steps between these two wins. Hence if $t > n_0$,

$$\ln(f(\boldsymbol{X}_t)) \leq \ln(f(\boldsymbol{X}_1)) - (t - n_0) \times$$
$$\sum_i \frac{\max(\ln(1/\eta), \ln(K'/K'') + \Delta_i)}{\min(1 + \Delta_i/\ln(2), 1 + \max_j \ln(\|\boldsymbol{\delta}_j\|)/\ln(2))} \tag{15}$$

where the summation is for i index of an iteration t with a "win", and n_0 is the number of initial iterations before a "win" (i.e. n_0 depends on the initial conditions but it is finite).

Equation 15 yields the expected result. $\square$

This result would be void if there was no algorithm and no space of functions for which assumptions 1–6 hold. Therefore, next Section is devoted to showing that for the important case of families of quadratic functions with bounded conditioning, assumptions 1–6 hold, and therefore the theorem above holds.

4   Application to Quadratic Functions

This section shows an example of application of the theorem above. The main strength of our results is that it covers many families of functions; yet, Eqs. 1–6 are not so readable. We show in this section that a simple family of fitness functions verify all the assumptions.

We consider the application to positive definite quadratic forms with bounded conditioning, i.e. we consider $f \in F$ with F the set of quadratic positive definite objective functions f such that

$$\frac{max\ Eigen\ Value(Hessian(f))}{min\ Eigen\ Value(Hessian(f))} < c_{max} < \infty. \tag{16}$$

Notably, thanks to the use of VC-dimension, the approach is indeed quite generic and can be applied to all families of functions obtained by rotation/translation from fitness functions in Fig. 1.

Instead of working on Q directly, with $x \mapsto Q(x - x^*)$ a quadratic form with Q positive definite with optimum in 0, we work on $x \mapsto \sqrt{Q(x - x^*)}$, so that Eq. 1 is verified; as considered algorithms are invariants by composition with monotone functions, this does not change the result.

We assume that $\delta_1, \delta_2, \ldots, \delta_k$ are independently uniformly randomly drawn in the unit ball $B(0,1)$. From now on, we note $p = p_{x,\sigma,f}$ the probability that $x + \sigma\delta_i$ is in $E = f^{-1}([0, f(x)[)$, and $\hat{p} = \hat{p}_{x,\sigma,f}$ the frequency $\frac{1}{k}\sum_{i=1}^{k}\chi_{x+\sigma\delta_i \in E}$. We will often drop the indices for short.

The assumptions in Sect. 2 essentially mean that frequencies are close to expectations for $x + \sigma\delta_i \in f^{-1}([0, f(x)[)$ and $x + \sigma\delta_i \in f^{-1}([0, \eta f(x)[)$, independently of x, σ, f. This is typically a case in which VC-dimension [14] can help.

The purpose of this section is to show Eqs. 1–6, for a given family of functions, namely the family F defined above; by proving Eqs. 1–6, we show the following.

Corollary: *Assume that the δ_i are uniformly randomly drawn in the unit ball $B(0,1)$. Assume that F is the set of quadratic functions with minimum in 0 ($f(0) = 0$) which verify Eq. 16 for some $c_{max} < \infty$. Then, almost surely on the sequence $\delta_1, \delta_2, \ldots, \delta_k$, for k large enough and some parameters k_1 and k_2 of Algorithm 1, then Eqs. 1–6 hold, and therefore for some $K < 0$, for all $t > 0$,*

$$\ln(\|X_t\|)/t \le K \tag{17}$$

with $\ln(0) = -\infty$ and where the sequence of X_t is defined as in Algorithm 1.

Proof: We use the main theorem above for proving Eq. 17, so we just have to prove that Eqs. 1–6 hold.

Step 1: Using VC-dimension for Approximating Expectations by Frequencies.
Thanks to the finiteness of the VC-dimension of quadratic forms (see e.g. [5]), we know that for all $\epsilon > 0$, almost surely in $\delta_1, \delta_2, \ldots, \delta_k$, for all $\delta > 0$ and k sufficiently large, with probability at least $1 - \delta$,

$$\sup_{x,f,\sigma>0} |\hat{p}_{x,\sigma,f} - p_{x,\sigma,f}| \leq \epsilon/2 \tag{18}$$

where x ranges over the domain, f ranges over F.

For short, we will often drop the indices, so that Eq. 18 becomes Eq. 19:

$$\sup_{x,f,\sigma>0} |\hat{p} - p| \leq \epsilon/2 \tag{19}$$

The important point here is that this result is a uniform result (uniform on $f \in F$); this is not just a simple law of large numbers, it is a uniform law of large numbers, so that it is not a mistake if there is a supremum on x, σ, f. Almost surely, the supremum is bounded; it is not only bounded almost surely for each x, σ, f separately, and this is the key concept in this proof.

Step 2: showing that σ small leads to high acceptance rate and σ high leads to small acceptance rate.
Thanks to the bounded conditioning (Eq. 16), there exists $\epsilon > 0$ s.t.

$$s' < \frac{1}{2}s \tag{20}$$

$$\text{with } s = \sup\left\{ \frac{\sigma}{||x||}; \sigma, x, f \text{ s.t. } p \geq \frac{\epsilon}{2} \right\}$$

$$\text{and } s' = \inf\left\{ \frac{\sigma}{||x||}; \sigma, x, f \text{ s.t. } p < \frac{1}{2} - \frac{\epsilon}{2} \right\}$$

because $s' \to 0$ and $s \to \infty$ as $\epsilon \to 0$.

Equation 18 implies

$$\frac{1}{2}\hat{s} \geq \frac{1}{2}s \tag{21}$$

and

$$s' \geq \hat{s}' \tag{22}$$

$$\text{with } \hat{s} = \sup\left\{ \frac{\sigma}{||x||}; \sigma, x, f \text{ s.t. } \hat{p} \geq \epsilon \right\}$$

$$\text{and } \hat{s}' = \inf\left\{ \frac{\sigma}{||x||}; \sigma, x, f \text{ s.t. } \hat{p} < \frac{1}{2} - \epsilon \right\}$$

So, Eqs. 19, 21 and 22, with k large enough, imply

$$\frac{1}{2}\hat{s} > \hat{s}' \tag{23}$$

Equation 23 provide k_1, k_2, c' and b' as follows for Eqs. 3 and 4:

$$\frac{1}{b'} = \hat{s} \;, \qquad \frac{1}{c'} = \hat{s}',$$

$$k_1 = \lfloor \epsilon k \rfloor \;\; \text{and} \;\; k_2 = \left\lceil (\frac{1}{2} - \epsilon)k \right\rceil$$

with $\epsilon < \frac{1}{10}$ (due to step 1). Equations above imply $c' \geq 2b'$.

Step 3: showing that k large enough and σ well chosen leads to at least one mutation with significant improvement. Similarly, k large enough yield

$$b^{-1} = \sup \left\{ \frac{\sigma}{||\boldsymbol{x}||}; \sigma, \boldsymbol{x}, f \text{ s.t. } \hat{p} > k_1/k \right\},$$

$$c^{-1} = \inf \left\{ \frac{\sigma}{||\boldsymbol{x}||}; \sigma, \boldsymbol{x}, f \text{ s.t. } \hat{p} < k_2/k \right\},$$

which provide Eqs. 5 and 2 with $b < c$ thanks to $\epsilon < \frac{1}{10}$ (ϵ was chosen with $\epsilon < \frac{1}{10}$ in step 1). Equations 2–5 then imply $b < b'$ and $c' < c$.

We now have to ensure Eq. 6. Equations 1–5 are proven above for k sufficiently large; from now on, we note $q = q_{\boldsymbol{x},\sigma,f}$ the probability that $\boldsymbol{x} + \sigma\boldsymbol{\delta}_i$ is in $E' = f^{-1}([0, \eta f(\boldsymbol{x})[)$, and $\hat{q} = \hat{q}_{\boldsymbol{x},\sigma,f}$ the frequency $\frac{1}{k}\sum_{i=1}^{k} 1_{\boldsymbol{x}+\sigma\boldsymbol{\delta}_i \in E'}$. For showing Eq. 6, let us assume

$$b^{-1} \leq \frac{\sigma}{||\boldsymbol{x}||} \leq c^{-1};$$

this implies $q > \epsilon_0$ for some $\epsilon_0 > 0$; so for k sufficiently large for ensuring $\sup_{\sigma,\boldsymbol{x},f} |q_{\boldsymbol{x},\sigma,f} - \hat{q}_{\boldsymbol{x},\sigma,f}| \leq \epsilon_0/2$, by VC-dimension, we get $q' \geq \epsilon_0/2 > 0$, which implies that at least one $\boldsymbol{\delta}_i$ verifies $\boldsymbol{x} + \boldsymbol{\delta}_i \in E'$. This is exactly Eq. 6.

Step 4: Concluding. We have shown Eqs. 1–6 for square roots of positive definite quadratic normal forms with bounded conditioning. Therefore, the main theorem can be applied and leads to Eq. 17.   □

5   Discussion and Conclusion

This work provides, to the best of our knowledge, the first proof of linear convergence of evolutionary algorithms (here, the Simple Evolution Strategy in Algorithm 1) in continuous domains on non quasi-convex functions. Indeed, even the application to quadratic positive definite forms is new. This proof is for derandomized samplings only, which means that the mutations δ_i, before multiplication by the step-size which obviously varies, are constant. A main missing point for an application is the evaluation of the convergence rate as a function of condition numbers (see extensions below) and the extension to randomized algorithms preferred by many practitioners.

In Sect. 5.1 we discuss extensions of this paper that we plane to develop in the near future, and in Sect. 5.2 deeper (harder to get rid of) limitations.

5.1 Extensions

Two properties are used for applying our main theorem to quadratic functions with a bound on condition numbers:

- VC-dimension of level sets. VC-dimension is a classical easy tool for showing that a family of functions verify a property such as Eq. 19 for arbitrarily small $\epsilon > 0$, provided that k is large enough.
- Equation 20, also crucial in the proof, is directly a consequence of bounded conditioning (assumption formalized in Eq. 16).

With these two assumptions, we can show Eqs. 1–6, and then the theorem can be applied. This is enough for objective functions with level sets having simple graphical representations with rotations/translations.

However, we do not need assumptions so strong as finite VC-dimension for showing Eqs. 1–6. Glivenko-Cantelli results are enough; and for this, finiteness of the bracketing covering numbers, for example, is enough [13]; this is the most natural extension of this work. In particular, there are results showing the finiteness of bracketing covering numbers for families of Hölderian spaces of functions; this is a nice path for applying results from this paper to wide families of functions.

Assumptions in [2] are slightly different from assumptions in this paper; their main assumption are

- the frontier of any level set $f^{-1}(r)$ has a bounded curvature.
- for some $C_{min} \in \mathbb{R}$ and $C_{max} \in \mathbb{R}$, with $\boldsymbol{x}^*$ the (assumed unique) optimum of the objective function f and $f(\boldsymbol{x}^*) = 0$, for any $r \in \mathbb{R}$, we have

$$B(\boldsymbol{x}^*, C_{min}r) \subset f^{-1}(r) \subset B(\boldsymbol{x}^*, C_{max}r).$$

The second assumption is equivalent to our conditioning assumption, but the first one is not directly equivalent to our derandomized sampling assumptions. Refining the assumptions might be possible by combining their assumptions and our assumptions.

Condition numbers are classical for estimating the difficulty of local convergence; a nice condition number for difficult optimization should generalize some classical condition number from the literature, and include non-differentiable functions as well. Reference [12] did a first step for that; in particular, isotropic algorithms do not solve functions with infinite condition number (for the definition of [12]), whereas covariance-based algorithms [8,10] do. Equation 7 we guess that it is possible to derive a new such number with direct links to convergence rates of evolution strategies.

5.2 Limitations

In this paper, we work on an evolutionary algorithm for which mutations δ_i's are randomly drawn once and for all (the same mutation vectors $\delta_1, \ldots, \delta_k$ for all iterations of the algorithms). This makes the proof much easier. We believe that the proof can be extended to the case in which the mutations are randomly drawn at each iteration, as in most usual cases; yet, the adaptation is not straightforward; we must study the frequency (over iterations) at which assumptions 2–6 hold, and the consequences of bad cases on Eq. 15. For this paper, we just assume that the δ_i's are randomly drawn once and for all iterations; equivalently, they could be quasi-randomized.

Cumulative adaptation [9] is not considered in our analysis; this is a considerably harder step for generalizing our results, because then the simple separation between 5 cases (see Fig. 2) is the idea that clearly divides the proof between step-size adaptation and progress rate.

This work covers quadratic functions, but the rates are not independent of the conditioning, so complementary results are necessary for algorithms evolving a covariance matrix, such as [3,8,10]. Maybe ergodic Markov chains are a better tool for showing such results [1].

We work under assumptions which imply a very large k. More precisely, using VC-dimension or bracketing numbers, it is possible to get explicit bounds on k, but these numbers will be far above the usual values for k. Obtaining results for limited values of k is a classical challenge in machine learning, and for the moment only huge values of k are applicable when using VC-dimension assumptions. Seemingly, weaker assumptions are enough, such as Glivenko-Cantelli properties [13]. For this paper, VC-dimension is easier to use and sufficient for our purpose.

Acknowledgements. We are grateful to Rémi Bergasse [2] for interesting discussions.

References

1. Auger, A.: Convergence results for (1,λ)-SA-ES using the theory of φ-irreducible Markov chains. Theor. Comput. Sci. **334**(1–3), 35–69 (2005)
2. Bergasse, R.: Stratégies d'évolution dérandomisées. Technical report, Ecole Normale Supérieure de Lyon (2007)
3. Beyer, H.-G., Sendhoff, B.: Covariance matrix adaptation revisited – The CMSA evolution strategy –. In: Rudolph, G., Jansen, T., Lucas, S., Poloni, C., Beume, N. (eds.) PPSN 2008. LNCS, vol. 5199, pp. 123–132. Springer, Heidelberg (2008)
4. Conn, A., Scheinberg, K., Toint, L.: Recent progress in unconstrained nonlinear optimization without derivatives. Math. Program. **79**(1–3), 397–414 (1997)
5. Devroye, L., Györfi, L., Lugosi, G.: A Probabilistic Theory of Pattern Recognition, vol. 31. Springer, New York (1997)
6. Droste, S., Jansen, T., Wegener, I.: On the optimization of unimodal functions with the (1 + 1) evolutionary algorithm. In: Eiben, A.E., Bäck, T., Schoenauer, M., Schwefel, H.-P. (eds.) PPSN 1998. LNCS, vol. 1498, p. 13. Springer, Heidelberg (1998)

7. Gould, N.I.M., Orban, D., Toint, P.L.: CUTEr and sifDec: A constrained and unconstrained testing environment, revisited. ACM Trans. Math. Softw. **29**(4), 373–394 (2003)
8. Hansen, N., Ostermeier, A.: Completely derandomized self-adaptation in evolution strategies. Evol. Comput. **9**(2), 159–195 (2001)
9. Ostermeier, A., Gawelczyk, A., Hansen, N.: Step-size adaptation based on non-local use of selection informatio. In: Davidor, Y., Männer, R., Schwefel, H.-P. (eds.) PPSN 1994. LNCS, vol. 866, pp. 189–198. Springer, Heidelberg (1994)
10. Schwefel, H.-P.: Numerical Optimization of Computer Models. Wiley, New York (1981). 1995–2nd edn.
11. Back, T., Hoffmeister, F., Schewefel, H.: A survey of evolution strategies. Technical Report, Department of Computer Science XI, University of Dortmund, D-4600, Dortmund 50, Germany (1991)
12. Teytaud, O.: Conditionning, halting criteria and choosing lambda. In: EA07, Tours France (2007)
13. van der Vaart, A., Wellner, J.: Weak Convergence and Empirical Processes: Springer series in statistics. Springer, New York (1996)
14. Vapnik, V., Chervonenkis, A.: On the uniform convergence of relative frequencies of events to their probabilities. Theory Probab. Appl. **16**, 264–280 (1971)

Ant Colony Optimization

Effective Multi-caste Ant Colony System for Large Dynamic Traveling Salesperson Problems

Leonor Melo[1,2](✉), Francisco Pereira[1,2], and Ernesto Costa[2]

[1] DEIS, ISEC, Instituto Politécnico de Coimbra, Rua Pedro Nunes,
3030-199 Quinta da Nora, Coimbra, Portugal
leonor@isec.pt
[2] Centro de Informática e Sistemas da Univ. Coimbra, 3030-790 Coimbra, Portugal
{xico,ernesto}@dei.uc.pt

Abstract. Multi-caste ant algorithms allow the coexistence of different search strategies, thereby enhancing search effectiveness in dynamic optimization situation. We present two new variants for a multi-caste ant colony system that promote a better migration of ants between alternative behaviors. Results obtained with large and highly dynamic traveling salesperson instances confirm the effectiveness and robustness of the approach. A detailed analysis reveals that one of the castes should adopt a clearly exploratory behavior, as this minimizes the recovery time after an environmental change.

Keywords: Ant colony optimization · Dynamic traveling salesperson problem · Multi-caste ant colony system · Traffic factor

1 Introduction

Ant Colony Optimization (ACO) encompasses a class of algorithms loosely inspired in the behavior of ants [4]. First developed to deal with the Traveling Salesperson Problem, it has proven successful in a wide range of hard combinatorial optimization problems [4]. Ant Colony System (ACS) [3] is one of the most successful ACO variants and its main distinguishing feature is the existence of a greedy decision rule adopted by artificial ants when building a solution for the problem being solved.

ACS, as well as other ACO variants, depend on a set of parameters that govern the way the search is conducted. Although beneficial, a careful adjustment of the settings is far from trivial. Also, the ideal setting may change throughout an optimization run, as the search conditions vary. In dynamic environments, where the problem modifies over time, this situation is amplified. In a previous work [15] we proposed a multi-caste framework that allows the coexistence of different sets of parameter values, hence search strategies, inside a single ACS algorithm. Also, although the total colony size is fixed, ants may migrate between castes

© Springer International Publishing Switzerland 2014
P. Legrand et al. (Eds.): EA 2013, LNCS 8752, pp. 67–78, 2014.
DOI: 10.1007/978-3-319-11683-9_6

during the run, thereby favoring the specific search strategy that seems to be more suitable at a given period. In [16], the multi-caste ACS was applied to several Dynamic Traveling Salesperson Problem (DTSP) instances. Results revealed that the adoption of different castes enhances the robustness of the algorithm, even though the absolute best performance was usually achieved by a standard ACS with an ideal fixed setting.

In this paper we extend the original framework by proposing two new multi-caste variants. The goal is to foster an efficient migration of ants between castes, thereby promoting a fast adaptation to the different scenarios that arise during search. Also, we test our framework with larger DTSP instances, with over 1000 cities. Dynamism is inserted by modifying the travel cost between cities. Several scenarios, differing in frequency and magnitude of change, are considered. Results show that the new multi-caste configurations are effective and outperform both standard ACS with ideal fixed settings and previous multi-caste variants, particularly in large DTSP instances.

The structure of the paper is the following: in Sect. 2 we present the multi-caste ACS used in our work. Section 3 describes the DTSP, whereas Sect. 4 comprises a presentation of the optimization results and corresponding analysis. Finally, Sect. 5 gathers the conclusions and suggests directions for future work.

2 Multi-caste Ant Colony System

While foraging, ants lay pheromones, a chemical signal, on the ground. This will gradually guide ants towards promising trails, thereby leading to the emergence of an indirect form of communication. Artificial ants belonging to ACO algorithms mimic this behavior and rely on an artificial trail to share information about the problem being solved. The optimization cycle of a general ACO method comprises two main steps: first, each ant builds a solution biased by pheromone values and specific heuristic information; afterwards, the pheromone values of the artificial trail are updated to reflect the quality of the new solutions found. This procedure is repeated until a termination criterion is satisfied.

2.1 Standard ACS

Ant System (AS) was the first ACO algorithm proposed [4]. Later on, ACS was presented [3] with the aim of improving AS effectiveness. ACS differs from previous variants in three key issues (see the aforementioned references for details): ants rely on a greedy decision rule to build the solutions, only the best ant is allowed to update the artificial pheromone trail and a local pheromone updating rule prevents the algorithm from excessive convergence. Parameter q_0 controls, in a probabilistic way, the amount of greediness the ants should use when constructing a solution. It is critical to the success of the ACS, as it balances the relative importance given to exploitation vs. exploration.

There are several studies in the literature that report the limitation of relying on ACS with fixed settings. In the static TSP, varying the parameter values as the search progresses might lead to an overall performance enhancement [9]. Several experiments reported in another study [15], reveal that the best parameter values also depend on the instance being optimized. As for the DTSP, the analysis described in [omitted reference] shows that distinct dynamic scenarios require ACS algorithms with different q_0 values.

2.2 Multi-caste ACS

In the multi-caste ACS, artificial ants are divided in several groups or castes. Each group encodes its own q_0 value, a parameter that strongly influences ACS search behavior. The idea is to grant ACS with different strategies, allowing it to select the best ant at any given search stage. Additionally, the alterations introduced in the conventional ACS are minimal. Ants inherit the setting from the caste to which they belong and, when applying the state transition rules, rely on their specific q_0 value. High q_0 castes contain exploitive ants, whereas lower q_0 castes are composed by more explorative ants, important to escape local optima and recover from a modification in the dynamic environment. All colonies start with the same number of ants and the total number of ants remains constant. The term caste, when applied artificial ants, appeared for the first time in [1], although no implementation was suggested.

When applying multi-caste ACS to a dynamic TSP we must ensure that, whenever a change occurs, the solution found by the best-so-far ant is re-evaluated using the new distance matrix. This is necessary since the same tour after the change could be associated with a bigger travel distance. Keeping the old, smaller, value would prevent algorithm from updating the best-so-far ant, thus a sub-optimal solution would be used to update the trail.

In [15] two multi-caste variants were proposed:

const-multi-caste The dimension of the castes is fixed throughout the optimization;

jump-multi-caste At the end of each iteration, two ants are selected at random. If the ants belong to different castes and both castes have more than 20 % of the total number of ants, the quality of their solutions is compared. The ant with the worse solution jumps to the caste of the winning ant. The idea behind this variant is to provide a simple method to dynamically adjust the size of the castes, favoring those that in the current search status encode the most promising q_0 value.

These two variants proved to be robust and able to adapt to different dynamic scenarios, particularly the jump-multi-caste. However, they tend to be outperformed by the conventional ACS with ideal settings. We hypothesize that this might be due to the waste of resources (i.e., ants) in sub-optimal castes: this happens either by keeping the caste's size constant (const-multi-case) or by an inefficient adjustment of the castes size when the optimization conditions change (jump-multi-caste). To allow for a better change in the dimension of the castes, i.e., to enforce a fast or more pronounced switch in the search strategy, we propose two new jump variants. At the end of each iteration, one ant from each caste is selected (independently of the castes size). If the ant with the worse solution comes from a caste with, at least, 2 ants, it jumps to the caste of the ant with the best solution. This simple rule allows a large concentration of ants in promising castes and, at the same time, it prevents extinction. There are two possible alternatives to select ants for the comparison:

super-jump-multi-caste One ant from each caste is selected at random.
greedy-jump-multi-caste The best ant from each caste is selected.

3   Dynamic Travelling Salesperson Problem

When considering DTSP, two types of dynamism can be devised: adding/removing cities to the problem or changing the cost between pairs of cities. Existing ACO techniques for the DTSP are usually aimed to one of the two variants.

In our work, we adopt the second possibility, also known as DTSP with a traffic factor [12]. For each pair of cities, i and j, $e_{ij} = d_{ij} \times f_{ij}$, where d_{ij} is the original distance between cities i and j, and f_{ij} is the traffic factor between those cities. Every F evaluations a random number R in $[F_L, F_U]$ is generated probabilistically. F_L and F_U are the upper and lower bounds for the traffic factor and R represents the traffic at that moment. With a probability M each link can change its traffic factor to $f_{ij} = 1 + R$, or otherwise, reset its traffic factor to 1 (meaning no traffic). F and M represent the number of iterations between changes (i.e., its frequency) and the magnitude of change, respectively.

To be able to compare the different algorithms, and since we know at which time step changes occur, we use the offline performance [2], that consists on the average of the best-since-last-change at each time step.

To determine which algorithm was more robust to change, i.e., had a smaller tour length immediately after a change we also measure the average peak, as described in formula 1.

$$Q = \frac{1}{E} \sum_{i \in C} \left(\frac{1}{T} \sum_{j=1}^{T} P_{ij} \right) \tag{1}$$

where C is the set of iterations immediately after a change, T is the number of independent runs, and P_{ij} is the best solution found at iteration i of run j.

3.1   Related Work

The existing ACO approaches for the DTSP cover both forms of dynamism: adding/removing cities to the problem [6–8,11,12,18] or changing the cost between pairs of cities (inserting traffic jams) [5,10,13,14]. Different dynamic scenarios have been considered by ACO algorithms. Some of the most relevant are:

– frequency of change: intervals between changes range from 20 iterations [11–14] to 750 iterations [7,8], or somewhere in between [18]; further, single or multiple changes can be considered [5,6,10];

– severity of change - from 0.5 % [7,8] to 75 % [11,12] of the cities added/removed; from 1 % [5,10] to 75 % [13,14] of the links affected by traffic jams;

DTSP instances selected to test the algorithms are usually of moderate size, varying between 25 ([5]) and 532 cities ([13]). No wide accepted benchmark exists.

The existing algorithms base their approaches mainly in: some sort of trail equalization or adjustment ([5,6,8,18]); a local search (KeepElitist) or other transformation procedure applied to an ant or a group of ants when a change occurs ([7,8,11,12,14,18]); explicit memory ([7,13,14,18]); an immigrants scheme ([11–14]). The approach described in [10] relies on a rank based, Q-learning inspired variant of ACO.

4 Experiments

We used the ACOTSP software [19], both to get the results for the standard ACS and as the base for our own implementations. Unless otherwise noted, the default values used for the experiments are: $m = 10$, $\beta = 2$, $\rho = 0.1$, $\xi = 0.1$, $\tau_0 = 1/(n \cdot L_{nn})$ (where L_{nn} is the length of the tour using the nearest neighbor heuristic [4,19]), and the local search algorithm is the 3-opt. In multi-caste configurations all castes have the same (initial) size. Different q_0 values, ranging from 0.1 to 0.99 were considered. Each experiment was repeated for 30 times.

Several TSP instances, with a number of cities between 532 and 1173, were selected from the TSPLIB 95 [17] to construct our dynamic scenarios, as described in Sect. 3. For every instance we consider 20 dynamic scenarios (4 values of $F \times$ 5 values of M): $F = \{10, 20, 100, 200\}$, where $F = 10$ defines a rapid changing environment and $F = 200$ represents a slow changing environment; $M = \{10, 25, 50, 75, 90\}$, with $M = 10$ and $M = 90$ establishing a small and large degree of change, respectively. For each instance and M value, 900 distance matrixes were created (30 distance matrixes per run × 30 runs). Each try was allowed to run for at least $30 \times F$ iterations.

For each dynamic scenario we compared each pair of configurations, using the paired t-test with confidence level of 0.95 and 29 degrees of freedom. Then, for each configuration c, we constructed a table (see Fig. 1). Let $v_{f,m}$ be the value present in the cell at line f and column m: it represents the comparative performance of the configuration c in a scenario where change occurs every f iterations and affects $m/100$ of the links, and is calculated as $v_{f,m} = \#w - \#b$, where w is the set of configurations that have an average offline performance statistically larger (worse) than c, and b is the set of configurations with a statistically smaller (better) offline performance than c. The bigger the value $v_{f,m}$ (the lighter the cell shading), the better that configuration performed when compared to the others, in that scenario. In the remainder of this section, we identify the ACS configurations according to the following convention:

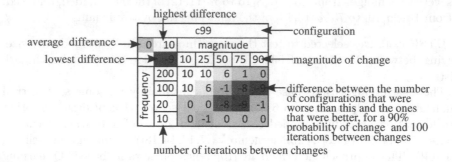

Fig. 1. Example of how a performance table should be read

Fig. 2 tables:

c99

3	18	magnitude			
-13	10	25	50	75	90
200	16	18	13	8	4
100	17	13	5	-9	-12
20	4	5	-13	-13	-1
10	1	-1	2	0	1

(frequency)

j01_99

-5	9	magnitude			
-15	10	25	50	75	90
200	-15	-15	-14	-14	-3
100	-14	-12	-13	4	0
20	-15	-4	9	6	9
10	1	0	1	2	-4

j05_99

-4	9	magnitude			
-15	10	25	50	75	90
200	-15	-15	-13	-11	-6
100	-14	-15	-12	-3	3
20	-3	9	7	6	3
10	0	0	1	2	2

j10_99

-6	4	magnitude			
-15	10	25	50	75	90
200	-15	-14	-14	-13	-7
100	-14	-13	-13	-2	-7
20	-3	-1	0	1	4
10	1	-2	-1	-2	1

j25_99

-7	5	magnitude			
-15	10	25	50	75	90
200	-15	-15	-13	-13	-11
100	-14	-15	-13	-4	-1
20	-7	-9	1	5	3
10	0	-5	-5	-2	-2

j50_99

-8	1	magnitude			
-14	10	25	50	75	90
200	-5	-9	-12	-13	-14
100	-14	-10	-13	-10	-9
20	-6	-8	0	-4	-4
10	-1	-3	1	-10	-11

j75_99

-9	3	magnitude			
-16	10	25	50	75	90
200	3	0	-10	-4	-13
100	-1	-10	-10	-15	-14
20	-11	-16	-13	-11	-12
10	-9	-4	-10	-8	-4

sj01_99

5	11	magnitude			
-6	10	25	50	75	90
200	-6	2	8	6	5
100	2	5	4	6	4
20	4	9	3	11	10
10	3	6	3	5	1

sj05_99

4	11	magnitude			
-2	10	25	50	75	90
200	0	5	1	5	5
100	6	5	4	5	10
20	6	9	11	6	5
10	-2	3	1	2	0

sj10_99

2	7	magnitude			
-7	10	25	50	75	90
200	0	-7	0	5	6
100	-1	0	5	2	4
20	2	6	4	1	2
10	4	-1	1	0	7

sj25_99

3	8	magnitude			
-3	10	25	50	75	90
200	3	4	6	4	2
100	8	2	6	2	0
20	8	0	-2	5	-3
10	0	0	2	2	1

sj50_99

1	6	magnitude			
-5	10	25	50	75	90
200	4	5	6	5	3
100	6	5	6	-2	-1
20	2	-2	-5	-1	-5
10	0	0	0	-2	-2

sj75_99

-0	14	magnitude			
-14	10	25	50	75	90
200	14	9	6	1	3
100	10	5	3	-2	-5
20	2	-10	-14	-8	-10
10	-2	-4	0	1	-4

gj01_99

5	9	magnitude			
0	10	25	50	75	90
200	1	3	8	6	3
100	1	8	5	9	4
20	3	5	3	7	7
10	1	5	0	4	7

gj05_99

5	15	magnitude			
-1	10	25	50	75	90
200	9	4	6	8	6
100	3	8	15	9	2
20	8	5	7	-1	0
10	1	-1	2	3	3

gj10_99

6	18	magnitude			
0	10	25	50	75	90
200	5	6	5	5	6
100	4	5	8	12	18
20	2	6	5	5	3
10	1	10	4	0	5

gj25_99

2	7	magnitude			
-1	10	25	50	75	90
200	0	5	5	2	2
100	4	7	5	2	3
20	6	-1	4	2	0
10	0	-1	0	0	1

gj50_99

2	8	magnitude			
-8	10	25	50	75	90
200	3	4	5	8	6
100	4	6	5	4	0
20	-1	4	4	-8	-3
10	1	2	0	1	-3

gj75_99

0	13	magnitude			
-11	10	25	50	75	90
200	13	10	7	5	3
100	7	6	3	-8	1
20	-1	-7	-11	-9	-8
10	0	-4	-2	2	1

Fig. 2. Comparative performance of the ACS variants for the att532 instance

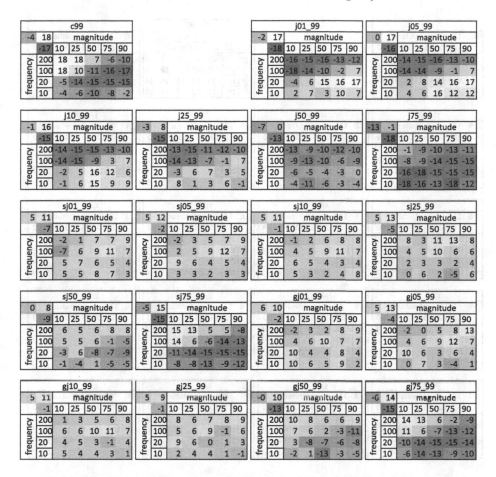

Fig. 3. Comparative performance of the ACS variants for the rat783 instance

- c99: standard ACS with $q_0 = 0.99$;
- jx_99 (sj$x$_99, gj$x$_99): jump (super-jump, greedy-jump) dual-caste configuration with q_0 values of $0.x$ and 0.99 (eg.: sj50_99).

Although other configurations were tested (namely const-multi-caste, and other q_0 values), due to space constraints and for the sake of clarity, we concentrate our analysis on the best performing configurations. A global overview of results can be consulted in Figs. 2, 3, and 4 for instances $att523$, $rat783$, and $pcb1173$ (results for other TSP instances follow the same trend). The outcomes clearly show the advantage of multi-caste variants over conventional ACS (please note that the standard ACS used in the analysis is already the best fixed configuration for the DTSP instances under study). The performance improvement is more evident, as the instances grow in size and in situations with a high degree of dynamism (big and/or frequent changes). This is an expected result, as insisting on the current trail when a considerable change just occurred (the typical

c99

-11	17	magnitude				
	-18	10	25	50	75	90
frequency	200	17	6	-17	-18	-18
	100	2	-16	-18	-18	-18
	20	-16	-15	-15	-15	-14
	10	-8	-8	-7	-14	-4

j01_99

7	17	magnitude				
	-15	10	25	50	75	90
frequency	200	-15	-12	-8	7	7
	100	-12	-2	8	12	16
	20	15	17	14	15	16
	10	11	13	7	15	15

j05_99

6	17	magnitude				
	-15	10	25	50	75	90
frequency	200	-15	-12	-7	2	7
	100	-15	-5	7	12	16
	20	13	12	15	16	17
	10	14	10	8	15	4

j10_99

5	16	magnitude				
	-18	10	25	50	75	90
frequency	200	-15	-18	-6	2	7
	100	-13	-3	9	14	16
	20	13	16	13	12	10
	10	8	7	7	6	10

j25_99

2	14	magnitude				
	-14	10	25	50	75	90
frequency	200	-14	-12	-9	-3	7
	100	-13	-7	7	0	6
	20	10	8	12	14	6
	10	6	5	7	2	9

j50_99

-5	3	magnitude				
	-12	10	25	50	75	90
frequency	200	-10	-12	-9	-7	-8
	100	-11	-8	-6	-3	-4
	20	0	-2	0	3	1
	10	-7	1	-11	-1	-9

j75_99

-14	-9	magnitude				
	-17	10	25	50	75	90
frequency	200	-9	-12	-11	-14	-14
	100	-13	-16	-14	-12	-12
	20	-13	-15	-15	-15	-15
	10	-13	-17	-13	-16	-17

sj01_99

6	11	magnitude				
	0	10	25	50	75	90
frequency	200	4	6	11	11	7
	100	9	10	9	5	5
	20	7	3	9	4	5
	10	8	0	7	3	2

sj05_99

7	13	magnitude				
	3	10	25	50	75	90
frequency	200	5	6	9	6	7
	100	10	13	7	8	6
	20	3	5	7	7	7
	10	8	7	7	8	3

sj10_99

6	11	magnitude				
	2	10	25	50	75	90
frequency	200	5	6	11	9	7
	100	7	10	2	9	5
	20	2	6	5	3	5
	10	6	6	4	3	8

sj25_99

3	14	magnitude				
	-3	10	25	50	75	90
frequency	200	5	4	14	8	7
	100	6	-3	7	-2	-1
	20	1	0	-3	1	-3
	10	2	-3	6	3	4

sj50_99

-5	6	magnitude				
	-12	10	25	50	75	90
frequency	200	5	6	-3	-5	-8
	100	6	4	-8	-9	-9
	20	-11	-6	-9	-8	-5
	10	-10	-7	-12	-2	-4

sj75_99

-11	6	magnitude				
	-15	10	25	50	75	90
frequency	200	5	6	-8	-14	-14
	100	5	-14	-14	-15	-15
	20	-15	-14	-15	-15	-14
	10	-13	-13	-13	-14	-13

gj01_99

6	12	magnitude				
	1	10	25	50	75	90
frequency	200	5	7	9	10	7
	100	6	12	7	10	6
	20	6	6	2	1	3
	10	4	4	7	11	3

gj05_99

6	13	magnitude				
	0	10	25	50	75	90
frequency	200	7	5	11	9	7
	100	8	13	7	11	5
	20	2	3	2	3	0
	10	5	5	7	2	3

gj10_99

4	10	magnitude				
	-3	10	25	50	75	90
frequency	200	5	8	9	8	7
	100	6	10	7	4	6
	20	3	2	4	3	4
	10	-3	-1	7	1	-2

gj25_99

2	11	magnitude				
	-10	10	25	50	75	90
frequency	200	5	6	11	8	7
	100	6	5	3	-2	-4
	20	3	-1	-4	-6	2
	10	-10	3	5	-2	-1

gj50_99

-5	6	magnitude				
	-11	10	25	50	75	90
frequency	200	5	6	2	-5	-8
	100	2	2	-6	-9	-9
	20	-11	-11	-7	-8	-10
	10	3	-3	-11	-10	-2

gj75_99

-10	6	magnitude				
	-15	10	25	50	75	90
frequency	200	5	6	-9	-14	-14
	100	4	-5	-14	-15	-15
	20	-12	-14	-15	-15	-15
	10	-11	-9	-12	-10	-9

Fig. 4. Comparative performance of the ACS variants for the pcb1173 instance

behavior of ACS with a high q_0), is not a good strategy. The problem is even more serious, if the time available to reach a new solution is limited.

Focusing our analysis on the three jump variants, it seems that the new super-jump and greedy-jump are more robust and effective than the original jump-multi-caste. The aggressive rules used to resize castes that were proposed in this paper lead to an efficient transfer of ants and allow for a faster recover when a change in the environment happens. The differences in performance between super-jump and greedy-jump are minimal in the tests performed and their behavior can be considered as equivalent. Overall the configurations gj01_99, gj05_99, gj10_99, sj01_99, and sj05_99 have consistently very good results. It is worth noting that the best multi-caste configurations keep a set of ants with an extremely low q_0. In concrete, 01_99 arrangements oscillate between two extremes, thereby adapting to different optimization scenarios: they can greedily exploit the best

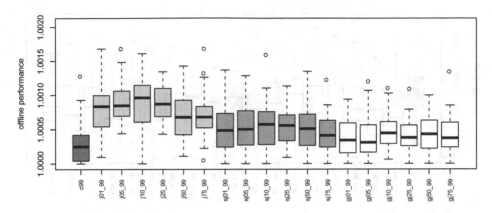

Fig. 5. Offline performance of the ACS variants on the att532 instance for the dynamic scenario $F = 100$ and $M = 25$.

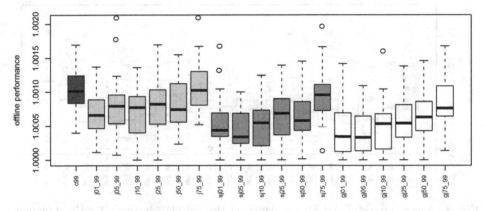

Fig. 6. Offline performance of the ACS variants on the pcb1173 instance for the dynamic scenario $F = 100$ and $M = 25$.

solutions found before, but, as soon as change happens, they quickly switch to an exploration mode.

Figures 5 and 6 display the offline performance of the 19 configurations previously considered in a moderate (att532) and large (pcb1173) DTSP instances, for a dynamic scenario obtained with $F = 100$ and $M = 25$ (the same trend is visible for other scenarios). A comparative analysis of the figures confirms our previous claims: the performance of the standard ACS deteriorates as the instances grow in size and it is clearly outperformed by nearly all multi-caste configurations in pcb1173. Also, within each variant, the relevance of specific q_0 values is more visible in the larger instance. The advantage of having a caste with a very low q_0 is obvious in Fig. 6.

Figures 7 and 8 show, for the same instances and dynamic scenarios, the normalized average offline performance (peak) measured in the iteration immediately

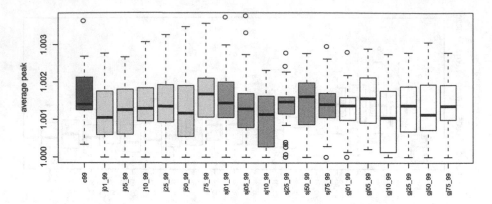

Fig. 7. Average peak of the ACS variants on the att532 instance for the dynamic scenario $F = 100$ and $M = 25$.

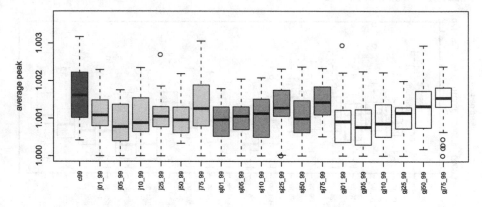

Fig. 8. Average peak of the ACS variants on the pcb1173 instance for the dynamic scenario $F = 100$ and $M = 25$.

after the change occurred (Eq. 1). As expected, conventional ACS has higher peaks immediately after change, as the ants are highly influenced by the current, possibly sub-optimal, trail. This holds true for most scenarios and instances. In the larger instance, it is evident that the presence of a caste with a low q_0 enhances the ability of avoiding extreme peaks after change.

5   Conclusions

Multi-caste ACO algorithms allow for the coexistence of different search strategies, thereby enhancing the plasticity and robustness of the method when solving a difficult optimization situation. In this paper we proposed two new multi-caste ACS variants that allow for a better migration of ants. This is particularly important in dynamic environments, where different scenarios appear over time. Results obtained with DTSP instances show that the new variants are effective and robust and outperform both the conventional ACS and previous multi-caste configurations. The advantage of the new proposals is more evident, when DTSP instances grow in size and when the degree of dynamism is higher. A noteworthy results is the need to maintain a caste with an extremely low q_0 value ($q_0 \leq 0.1$), that supervises the fast recovery of solutions when the environment changes.

All the tests reported in this paper dealt with dual-castes configuration. In the near future, we aim to study the advantages and weaknesses of generalizing our approach to a higher number of castes. Also, we intend to test this framework in different dynamic optimization problems.

References

1. Botee, H., Bonabeau, E.: Evolving ant colony optimization. Adv. Complex Syst. 1(2–3), 149–159 (1998)
2. Branke, J.: Evolutionary Optimization in Dynamic Environments. Kluwer Academic Publishers, Dordrecht (2002)
3. Dorigo, M., Gambardella, L.M.: Ant colony system: a cooperative learning approach to the traveling salesman problem. IEEE Trans. Evolut. Comput. 1(1), 53–66 (1997)
4. Dorigo, M., Stützle, T.: Ant Colony Optimization. A Bradford Book. MIT Press, Cambridge (2004)
5. Eyckelhof, C.J., Snoek, M.: Ant systems for a dynamic TSP: ants caught in a traffic jam. In: Dorigo, M., Di Caro, G.A., Sampels, M. (eds.) Ant Algorithms 2002. LNCS, vol. 2463, pp. 88–99. Springer, Heidelberg (2002)
6. Guntsch, M., Middendorf, M.: Pheromone modification strategies for ant algorithms applied to dynamic TSP. In: Boers, E.J.W., Gottlieb, J., Lanzi, P.L., Smith, R.E., Cagnoni, S., Hart, E., Raidl, G.R., Tijink, H. (eds.) EvoIASP 2001, EvoWorkshops 2001, EvoFlight 2001, EvoSTIM 2001, EvoCOP 2001, and EvoLearn 2001. LNCS, vol. 2037, pp. 213–222. Springer, Heidelberg (2001)
7. Guntsch, M., Middendorf, M.: Applying population based ACO to dynamic optimization problems. In: Dorigo, M., Di Caro, G.A., Sampels, M. (eds.) Ant Algorithms 2002. LNCS, vol. 2463, pp. 111–122. Springer, Heidelberg (2002)
8. Guntsch, M., Middendorf, M., Schmeck, H.: An ant colony optimization approach to dynamic TSP. In: GECCO'01 Proceedings of the Genetic and Evolutionary Computation Conference, pp. 860–867. Morgan Kaufmann Publishers (2001)
9. Hao, Z.F., Cai, R.C., Huang, H.: An adaptive parameter control strategy for ACO. In: Fifth International Conference on Machine Learning and Cybernetics. IEEE Press (2006)
10. Liu, J.: Rank-based ant colony optimization applied to dynamic traveling salesman problems. Eng. Optim. 37(8), 831–847 (2005)
11. Mavrovouniotis, M., Yang, S.: Ant colony optimization with immigrants schemes in dynamic environments. In: Schaefer, R., Cotta, C., Kołodziej, J., Rudolph, G (eds.) PPSN XI. LNCS, vol. 6239, pp. 371–380. Springer, Heidelberg (2010)
12. Mavrovouniotis, M., Yang, S.: A memetic ant colony optimization algorithm for the dynamic travelling salesman problem. Soft Comput. 15(7), 1405–1425 (2011)
13. Mavrovouniotis, M., Yang, S.: An immigrants scheme based on environmental information for ant colony optimization for the dynamic travelling salesman problem. In: Hao, J.-K., Legrand, P., Collet, P., Monmarché, N., Lutton, E., Schoenauer, M. (eds.) EA 2011. LNCS, vol. 7401, pp. 1–12. Springer, Heidelberg (2012)

14. Mavrovouniotis, M., Yang, S.: Memory-based immigrants for ant colony optimization in changing environments. In: Di Chio, C., Cagnoni, S., Cotta, C., Ebner, M., Ekárt, A., Esparcia-Alcázar, A.I., Merelo, J.J., Neri, F., Preuss, M., Richter, H., Togelius, J., Yannakakis, G.N. (eds.) EvoApplications 2011, Part I. LNCS, vol. 6624, pp. 324–333. Springer, Heidelberg (2011)
15. Melo, L., Pereira, F., Costa, E.: Multi-caste ant colony optimization algorithms. In: Local Proceedings of EPIA 2011, Lisbon, pp. 978–989 (2011)
16. Melo, L., Pereira, F., Costa, E.: Multi-caste ant colony algorithm for the dynamic traveling salesperson problem. In: Tomassini, M., Antonioni, A., Daolio, F., Buesser, P. (eds.) ICANNGA 2013. LNCS, vol. 7824, pp. 179–188. Springer, Heidelberg (2013)
17. Reinhelt, G.: {TSPLIB}: a library of sample instances for the TSP (and related problems). http://comopt.ifi.uni-heidelberg.de/software/TSPLIB95/
18. Sammoud, O., Solnon, C., Ghedira, K.: A new ACO approach for solving dynamic problems. In: Artificial Evolution (EA'09) - Local Proceedings (2009)
19. Stützle, T.: {ACOTSP}: software package of various ACO algorithms applied to the symmetric TSP (2002). http://www.aco-metaheuristic.org/aco-code/

Beam-ACO for the Repetition-Free Longest Common Subsequence Problem

Christian Blum[1,2]([✉]), Maria J. Blesa[3], and Borja Calvo[1]

[1] Department of Computer Science and Artificial Intelligence,
University of the Basque Country, San Sebastian, Spain
borja.calvo@ehu.es

[2] IKERBASQUE, Basque Foundation for Science, Bilbao, Spain
christian.blum@ehu.es

[3] ALBCOM Research Group, Universitat Politécnica de Catalunya, Barcelona, Spain
mjblesa@lsi.upc.edu

Abstract. In this paper we propose a Beam-ACO approach for a combinatorial optimization problem known as the repetition-free longest common subsequence problem. Given two input sequences x and y over a
finite alphabet Σ, this problem concerns to find a longest common subsequence of x and y in which no letter is repeated. Beam-ACO algorithms
are combinations between the metaheuristic ant colony optimization and
a deterministic tree search technique called beam search. The algorithm
that we present is an adaptation of a previously published Beam-ACO
algorithm for the classical longest common subsequence problem. The
results of the proposed algorithm outperform existing heuristics from
the literature.

1 Introduction

The classical longest common subsequence (LCS) problem is a string problem
in which a problem instance (S, Σ) consists of a set $S = \{s_1, s_2, \ldots, s_n\}$ of n
input strings over a finite alphabet Σ [7]. The problem is then about finding
a string being (1) as long as possible and (2) a subsequence of all the strings
in S. In this context, a string t is called a subsequence of a string s, if t can
be produced from s by deleting zero or more characters. For example, dga is a
subsequence of $adagtta$. A string that has both properties as described above is
called a *longest common subsequence* of the strings in S. The LCS problem has
applications in traditional computer science fields (such as data compression and
file comparison [2,13]) but also, for example, in computational biology [8,12].
Moreover, the LCS problem was shown to be NP-hard [10] for an arbitrary
number n of input strings.

The problem that we tackle in this work is a restricted version of the LCS problem, the so-called *repetition-free longest common subsequence (RFLCS)* problem.
Given exactly two input strings x and y over a finite alphabet Σ, the goal is to find
a longest common subsequence with the additional restriction that no letter may

© Springer International Publishing Switzerland 2014
P. Legrand et al. (Eds.): EA 2013, LNCS 8752, pp. 79–90, 2014.
DOI: 10.1007/978-3-319-11683-9_7

appear more than once. This problem was introduced in [1] for the purpose of having a comparison measure for two sequences of biological origin. In the same paper, the authors proposed three heuristics for solving this problem. These heuristics are up to now the only published algorithms for the RFLCS problem in the literature. Other variants of the classical LCS problem were studied, for example, in [5].

In contrast to the RFLCS, the classical LCS problem has already been subject of a multitude of research works over the past decades. Apart from algorithms based on dynamic programming and deterministic heuristics, the LCS has also attracted researchers from the field of metaheuristics (see [6,9,11]). The lastest one of the metaheuristic approaches for the LCS problem is Beam-ACO [3], a metaheuristic approach which results from a combination of ant colony optimization (ACO) with beam search (BS). In this work we adapt this Beam-ACO approach to the RFLCS problem and show that the performance of the resulting algorithm is mostly superior to the performance of the heuristics from the literature.

The organization of this paper is as follows. The proposed Beam-ACO approach is described in Sect. 2. Section 3 outlines the experimental evaluation. Finally, in Sect. 4 conclusions and an outlook to future work are offered.

2   Beam-ACO

In the following we first describe the ACO-based framework of the proposed algorithm. Afterwards, the BS component is presented. Note that the algorithmic framework is exactly the same as the one described in [3]. The adaptation of the algorithm from the LCS to the RFLCS problem concerns the BS component.

Data Structures and Pheromone Model. The type of ACO algorithm which was chosen for the algorithmic framework is a $\mathcal{MAX}$-$\mathcal{MIN}$ Ant System implemented in the hyper-cube framework (HCF) [4]. The pseudo-code is shown in Algorithm 1. The algorithm requires the following data structures: (1) the *best-so-far* solution T^{bs}, i.e., the best solution generated by the algorithm over time; (2) the *restart-best* solution T^{rb}, i.e., the best solution generated since the algorithm's last restart; (3) the *convergence factor* cf, $0 \leq cf \leq 1$, which is a measure indicating the state of the convergence of the algorithm; and (4) the Boolean control variable bs_update, which assumes value *true* when the algorithm reaches convergence.

One of the crucial components of any ACO algorithm is the pheromone model $\mathcal{T}$. In the context of this paper, $\mathcal{T}$ consists of a pheromone value $0 \leq \tau_{x,i} \leq 1$ for each position i of input sequence x ($i \in \{1, \ldots, |x|\}$), and a pheromone value $\tau_{y,j}$ for each position j of input sequence y ($j \in \{1, \ldots, |y|\}$). Observe that a pheromone value $\tau_{x,i}$ (respectively $\tau_{y,j}$) indicates the *desirability* of adding the letter at position i of string x (respectively, the letter at position j of string y) to the solution under construction.

This pheromone model allows to represent solutions to the problem (that is, repetition-free common subsequences) in a specific way. Note that any common subsequence t of strings x and y can be translated in a well-defined way into a

---

Algorithm 1. Beam-ACO for the RFLCS problem

1: **input:** $x, y, k_{\text{bw}}, \mu \in \mathbb{Z}^+$
2: $T^{bs} :=$ NULL, $T^{rb} :=$ NULL, $cf := 0$, $bs\_update :=$ FALSE
3: Initialize all pheromone values to 0.5
4: **while** CPU time limit not reached **do**
5:    $T^{pbs} :=$ ProbabilisticBeamSearch(k_{bw}, μ) {see Alg. 2}
6:    **if** $|t^{pbs}| > |t^{rb}|$ **then** $T^{rb} := T^{pbs}$
7:    **if** $|t^{pbs}| > |t^{bs}|$ **then** $T^{bs} := T^{pbs}$
8:    ApplyPheromoneUpdate($cf, bs\_update, \mathcal{T}, T^{pbs}, T^{rb}, T^{bs}$)
9:    $cf :=$ ComputeConvergenceFactor($\mathcal{T}$)
10:    **if** $cf > 0.99$ **then**
11:       **if** $bs\_update =$ TRUE **then**
12:          Re-init. all pheromone values to 0.5, $T^{rb} :=$ NULL, $bs\_update :=$ FALSE
13:       **else**
14:          $bs\_update :=$ TRUE
15:       **end if**
16:    **end if**
17: **end while**
18: **output:** the string version t^{bs} of T^{bs}

---

unique ACO-solution $T = (X, Y)$, where both X and Y are binary strings and X is of length $|x|$ while Y is of length $|y|$. Hereby, the meaning of $X[i] = 1$ is that the letter at position i of string x (that is, $x[i]$) was chosen for the construction of solution t, while $X[i] = 0$ means that $x[i]$ was not chosen. The same holds, for example, for $Y[j] = 1$, respectively $Y[j] = 0$. A solution t is translated into $T = (X, Y)$ as follows: first, the position of the left-most occurrence of $t[1]$ in x (where $t[1]$ is the first character of t) is determined, say k_1. Then, all $X[i]$ with $i < k_1$ are set to 0, while $X[k_1] := 1$. Next, the position of the first occurrence of $t[2]$ in x after position k_1 is determined, say k_2. Then, all $X[i]$ with $k_1 < i < k_2$ are set to 0, while $X[k_2] := 1$. This is continued until all positions of t are treated. Afterwards, the same procedure is applied to string y in order to produce Y.

Algorithmic Framework. The algorithm works as follows. First all pheromone values are initialized to 0.5. Then, at each iteration, a probabilistic version of BS based on pheromone values is applied. For a description of the BS component see Sect. 2.1. BS generates a solution T^{pbs} as output. Afterwards, a pheromone update is performed in ApplyPheromoneUpdate($cf, bs\_update, \mathcal{T}, T^{pbs}, T^{rb}, T^{bs}$). Moreover, the current value of the convergence factor cf is determined. Depending on cf and the value of the Boolean variable bs_update, a decision on whether to restart the algorithm is made. In case of a restart, all pheromone values are readjusted to 0.5. The stopping criterion of the algorithm is a maximum computation time. As output upon termination, the algorithm provides the string version t^{bs} of the best-so-far ACO-solution T^{bs}. The two remaining procedures of Algorithm 1 are detailed in the following.

ApplyPheromoneUpdate($cf, bs\_update, \mathcal{T}, T^{pbs}, T^{rb}, T^{bs}$): As a standard procedure, three solutions are used for updating the pheromone values: T^{pbs}, as generated

Table 1. Setting of κ_{pbs}, κ_{rb}, κ_{bs}, and ρ depending on the convergence factor cf and the Boolean control variable bs_update

	bs_update = FALSE				bs_update = TRUE
	$cf < 0.4$	$cf \in [0.4, 0.6)$	$cf \in [0.6, 0.8)$	$cf \geq 0.8$	
κ_{pbs}	1	2/3	1/3	0	0
κ_{rb}	0	1/3	2/3	1	0
κ_{bs}	0	0	0	0	1
ρ	0.2	0.2	0.2	0.15	0.15

by BS in the current iteration, T^{rb}, and T^{bs}. The weight of each solution for the purpose of the pheromone update is determined as a function of cf, the convergence factor. The pheromone values $\tau_{x,i}$ corresponding to input string x are updated as follows:

$$\tau_{x,i} := \tau_{x,i} + \rho \cdot (\xi_{x,i} - \tau_{x,i}), \tag{1}$$

where

$$\xi_{x,i} := \kappa_{pbs} \cdot X^{pbs}[i] + \kappa_{rb} \cdot X^{rb}[i] + \kappa_{bs} \cdot X^{bs}[i], \tag{2}$$

where κ_{pbs} is the weight of solution $T^{pbs} = (X^{pbs}, Y^{pbs})$, κ_{rb} the one of $T^{rb} = (X^{rb}, Y^{rb})$, κ_{bs} the one of $T^{bs} = (X^{bs}, Y^{bs})$, and $\kappa_{pbs} + \kappa_{rb} + \kappa_{bs} = 1$. The weight values that we chose are the standard ones shown in Table 1. Also, note that the same pheromone update rule as described above is applied to the pheromone values $\tau_{y,j}$ corresponding to input string y. Finally, note that the algorithm works with upper and lower bounds for the pheromone values, that is, $\tau_{max} = 0.999$ and $\tau_{min} = 0.001$. In case a pheromone values surpasses one of these limits, the value is set to the corresponding limit. This has the effect that a complete convergence of the algorithm is avoided.

ComputeConvergenceFactor($\mathcal{T}$): The formula that was used for computing the value of the convergence factor is as follows:

$$cf := 2 \left(\left(\frac{\sum_{\tau \in \mathcal{T}} \max\{\tau_{max} - \tau, \tau - \tau_{min}\}}{|\mathcal{T}| \cdot (\tau_{max} - \tau_{min})} \right) - 0.5 \right)$$

This implies that at the start of the algorithm (or after a restart) cf has value zero. On the other side, in the case in which all pheromone values are either at τ_{min} or at τ_{max}, cf has a value of one. In general, cf moves in $[0,1]$.

2.1　BS Component

The probabilistic BS component which is applied in Procedure ProbabilisticBeamSearch(k_{bw}, μ) of Algorithm 1 works as follows (see also the pseudo-code in Algorithm 2). Solutions are constructed from left to right, and partial solutions are

---

Algorithm 2. Procedure ProbabilisticBeamSearch(k_{bw},μ) of Algorithm 1

1: **input:** $x, y, k_{\mathrm{bw}}, \mu$
2: $B_{\mathrm{compl}} := \emptyset$, $B := \{\emptyset\}$, $t_{\mathrm{bsf}} := \emptyset$
3: **while** $B \neq \emptyset$ **do**
4:     $E_B :=$ Produce_Extensions(B)
5:     $E_B :=$ Filter_Extensions(E_B)
6:     $B := \emptyset$
7:     **for** $k = 1, \ldots, \min\{\lfloor \mu k_{\mathrm{bw}} \rfloor, |E_B|\}$ **do**
8:         $za :=$ Choose_Extension(E_B)
9:         $t := za$
10:         **if** UB(t) = $|t|$ **then**
11:             $B_{\mathrm{compl}} := B_{\mathrm{compl}} \cup \{t\}$
12:             **if** $|t| > |t_{\mathrm{bsf}}|$ **then** $t_{\mathrm{bsf}} := t$ **end if**
13:         **else**
14:             **if** UB(t) $\geq |t_{\mathrm{bsf}}|$ **then** $B := B \cup \{t\}$ **end if**
15:         **end if**
16:         $E_B := E_B \setminus \{t\}$
17:     **end for**
18:     $B :=$ Reduce(B, k_{bw})
19: **end while**
20: **output:** The ACO-version T^{pbs} of argmax $\{|t| \mid t \in B_{\mathrm{compl}}\}$

---

extended by appending exactly one letter at a time. The two input parameters of BS are $k_{\mathrm{bw}} \in \mathbb{Z}^+$, which is the so-called *beam width*, and $\mu \in \mathbb{R}^+ \geq 1$, which is a parameter used to determine the maximal number of solution extensions that may be chosen at each step. The algorithm maintains a set B (the *beam*) for storing the current set of partial solutions. At the start B is initialized with the empty string denoted by $\emptyset$. Let E_B denote the set of all possible extensions of the partial solutions in B. At each step, $\lfloor \mu k_{\mathrm{bw}} \rfloor$ of these extensions are selected based on a greedy function and the pheromone values. Hereby, complete (that is, non-extensible) solutions are stored in B_{compl}, and partial solutions are added to set B in case the corresponding upper bound value (as computed by function UB()) is greater than the length of the best-so-far solution t_{bsf}. In order to finalize a step, B must be reduced in case it contains more than k_{bw} partial solutions. This is done on the basis of the upper bound values. More specifically, the best partial solutions with respect to the upper bound values remain in B. In the following the four different procedures of Algorithm 2 are outlined in detail.

Produce_Extensions(B): Given the current beam B as input, this procedure generates a set E_B of non-dominated extensions of all the partial solutions in B, which is done as explained in the following. First, given a partial solution t, the reduced alphabet Σ^t only contains letters which do not appear in t. Furthermore, let $x = x^+ \cdot x^-$ be the partition of input sequence x into substrings x^+ and x^- such that t is a subsequence of x^+, and x^- has maximal length. In the same way, y^+ and y^- are defined. Given this partition, which is well-defined, *position pointers* $p_x := |x^+|$ and $p_y := |y^+|$ are introduced. Moreover, the position of the first appearance of a letter $a \in \Sigma^t$ in strings x and y after the position pointers p_x and p_y is well-defined and denoted by p_x^a and p_y^a. In case letter $a \in \Sigma^t$ does not appear in x (respectively y), p_x^a (respectively p_y^a) is set to ∞. In this context, a letter $a \in \Sigma^t$ is called *dominated*, if there exists at least one letter $b \in \Sigma^t$, $a \neq b$, such that $p_x^b < p_x^a$ and $p_y^b < p_y^a$. Finally, $\Sigma_{nd}^t \subseteq \Sigma^t$ denotes the set of non-dominated letters of the reduced alphabet Σ^t with respect to partial solution t. Observe also that letters in Σ_t^{nd} are required to appear at least once in both x^- and y^-. Finally, set E_B is generated as the set of subsequences ta, where $t \in B$ and $a \in \Sigma_{nd}^t$.

Filter_Extensions(E_B): The non-domination relation—as defined above—can also be considered for extensions of different partial solutions of the same length. Formally, given two extensions $ta, zb \in E_B$, where $t \neq z$ but not necessarily $a \neq b$, ta is said to dominate zb if and only if the position pointers concerning a appear before the position pointers concerning b in the corresponding remaining parts of the two input strings. Using this relation, E_B is filtered in order to remove all dominated elements.

Choose_Extension(E_B): This procedure handles the probabilistic choice of a partial solution from E_B, both on the basis of a greedy function and the pheromone values. The greedy value of an extension $ta \in E_B$ is computed as follows:

$$\eta(ta) := \left(\frac{p_x^a - p_x}{|x^-|} + \frac{p_y^a - p_y}{|y^-|} \right)^{-1} \qquad (3)$$

Instead of directly using these greedy values, we decided to use the corresponding ranks instead. More specifically, the final greedy value $\nu(ta)$ of a partial solution $ta \in E_B$ is calculated as the sum of the ranks of the greedy weights that correspond to the construction steps that were performed to construct string ta. With this definition of $\nu()$, the probability for each $ta \in E_B$ is computed as follows:

$$\mathbf{p}(ta|E_B) = \frac{\left(\min\{\tau_{x,p_x^a}, \tau_{y,p_y^a}\} \cdot \nu(ta)^{-1} \right)}{\sum\limits_{zb \in E_B} \left(\min\{\tau_{x,p_x^b}, \tau_{y,p_y^b}\} \cdot \nu(zb)^{-1} \right)} \qquad (4)$$

Remember that p_x^a was defined as the next position of letter a after position pointer p_x in string x, and similarly for p_y^a. The intuition of this formula is as follows: If at least one of the pheromone values τ_{x,p_x^a} and τ_{y,p_y^a} is low, the corresponding letter should not yet be appended to the string, because there seems to be another letter that should be appended first. Finally, each application of function Choose_Extension(E_B) is either executed probabilistically, or deterministically (by choosing the option with the highest probability). The probability for a deterministic choice, also called the *determinism rate*, is henceforth denoted by $q \in [0, 1]$.

Reduce(B, k_{bw}): This procedure reduces B, if necessary, to exactly k_{bw} elements, based on their upper bound value. Given a partial solution $t \in B$, $\delta(x, a)$ (for all $a \in \Sigma^t$) evaluates to one, in case letter a appears at least once in x^-. Otherwise,

$\delta(x, a)$ evaluates to zero. The same holds for $\delta(y, a)$. The upper bound value of $t \in B$ is then defined as follows:

$$\text{UB}(t) := |t| + \sum_{a \in \Sigma^t} \min \{\delta(x, a), \delta(y, a)\} \tag{5}$$

Note that this upper bound function can be efficiently computed by keeping appropriate data structures.

3 Experimental Evaluation

The experimental evaluation has been performed on a PC with an Intel i7 quad core processor with 3 GHz and 8 GB of memory. First, we re-implemented the two best heuristics for the RFLCS problem presented in [1]. These heuristics are henceforth labeled A1 and A2, just like in the original paper. A1 is a deterministic heuristic which computes a longest common subsequence t (using dynamic programming) of the input sequences x and y. Afterwards, all repetitions of letters in t are deleted, maintaining of each letter exactly one occurrence. A2 is a probabilistic heuristic which works as follows. Let $n(x, a)$ denote the number of occurrences of a letter a in string x. Moreover, let $m_a(x, y)$ be defined as $\min\{n(x, a), n(y, a)\}$. For each $a \in \Sigma$, if $m_a(x, y) = n(x, a)$ heuristic A2 picks uniformly at random one occurrence of a in x. All other occurrences of a in x are deleted. Otherwise, if $m_a(x, y) = n(y, a)$ heuristic A2 picks uniformly at random one occurrence of a in y. Again, all other occurrences of a in y are deleted. This results in sequences x' and y'. Finally, A2 computes a longest common subsequence of x' and y' and provides the result as output. As in the original paper, A2 was applied 20 times to each problem instance, and the best result was taken as the final result. Beam-ACO was applied once to each problem instance (remember that results are averaged over 10 problem instances), with a computation time limit of 5 CPU seconds per run, a beam width of 10, and a determinism rate of $q = 0.9$. Note that the short computation time and the standard parameter setting (without a tuning process) was chosen on purpose in order to show that even an un-tuned Beam-ACO with a short running time is able to outperform the existing heuristics.

Problem Instances. Two sets of problem instances were generated, following the procedure as described in [1]. The first set (henceforth called Set1) consists for each combination of input sequence length $n \in \{32, 64, 128, 256, 512\}$ and alphabet size $|\Sigma| \in \{n/8, n/4, 3n/8, n/2, 5n/8, 3n/4, 7n/8\}$ of exactly 10 problem instances. The second set of instances (henceforth called Set2) is generated on the basis of the alphabet size $|\Sigma| \in \{4, 8, 16, 32, 64\}$ and the maximal repetition of each letter $rep \in \{3, 4, 5, 6, 7, 8\}$ in each input string. For each combination of $|\Sigma|$ and rep this instance set consists of 10 randomly generated problem instances.

Table 2. Experimental results concerning the instances of Set1.

$|\Sigma|$	n	Heuristic A1			Heuristic A2			Beam-ACO		
		result	std	time (s)	result	std	time (s)	result	std	time (s)
	32	4.0	0.00	< 0.001	4.0	0.0	< 0.001	4.0	0.0	< 0.001
	64	7.8	0.42	< 0.001	8.0	0.0	< 0.001	8.0	0.0	< 0.001
$n/8$	128	15.1	0.74	< 0.001	15.4	0.52	0.0016	16.0	0.0	0.0036
	256	28.5	1.65	< 0.001	25.7	0.67	0.002	31.9	0.32	0.026
	512	51.9	1.45	0.0028	40.7	1.25	0.01	62.3	0.82	1.78
	32	6.8	0.63	< 0.001	7.6	0.52	< 0.001	7.9	0.32	< 0.001
	64	12.4	0.84	< 0.001	12.8	0.79	< 0.001	14.3	1.34	0.01
$n/4$	128	21.5	0.97	< 0.001	20.3	1.06	0.002	25.3	0.48	0.21
	256	35.2	2.15	< 0.001	30.3	1.64	0.0032	42.4	1.43	0.72
	512	59.0	4.03	0.0012	45.5	1.96	0.01	68.0	3.13	0.78
	32	7.3	0.48	< 0.001	8.6	0.84	< 0.001	8.7	0.68	< 0.001
	64	13.1	2.02	< 0.001	13.3	1.25	0.0012	14.4	1.17	0.0036
$3n/8$	128	22.1	2.56	< 0.001	21.2	1.87	0.002	25.1	2.13	0.063
	256	35.9	2.47	< 0.001	31.3	1.42	0.0052	39.7	2.31	0.24
	512	53.7	1.25	0.0024	42.6	2.22	0.002	59.4	1.84	1.31
	32	7.6	1.65	< 0.001	8.3	1.34	< 0.001	8.8	1.55	< 0.001
	64	13.2	1.92	< 0.001	13.8	1.14	0.0012	14.5	1.08	0.046
$n/2$	128	21.9	1.20	< 0.001	21.0	0.94	0.0024	23.4	0.97	0.048
	256	31.9	2.69	< 0.001	29.8	1.81	0.0044	34.1	2.28	0.17
	512	49.8	2.25	0.0016	43.7	2.26	0.0124	53.1	3.14	0.588
	32	7.4	0.97	< 0.001	7.9	0.88	< 0.001	7.9	0.88	< 0.001
	64	12.6	2.01	< 0.001	12.9	1.45	0.0012	13.7	1.64	0.0032
$5n/8$	128	19.6	2.59	< 0.001	19.5	1.58	0.0016	21.1	1.91	0.0156
	256	29.7	2.63	< 0.001	28.7	1.83	0.0068	31.1	2.73	0.15
	512	45.8	2.66	0.002	42.0	1.15	0.014	47.8	1.93	0.328
	32	7.0	1.25	< 0.001	7.7	0.95	< 0.001	7.8	1.14	< 0.001
	64	12.2	1.03	< 0.001	13.0	0.94	< 0.001	13.1	0.74	0.0048
$3n/4$	128	18.7	1.84	< 0.001	18.5	1.43	0.0028	19.1	1.97	0.0088
	256	29.0	2.00	< 0.001	28.5	1.96	0.004	30.0	1.94	0.06
	512	43.6	2.12	0.002	41.9	1.60	0.015	44.7	1.77	0.47
	32	7.1	1.20	< 0.001	7.5	1.51	< 0.001	7.6	1.58	< 0.001
	64	12.1	2.23	< 0.001	11.9	2.28	0.0012	12.2	2.15	0.002
$7n/8$	128	18.0	1.94	< 0.001	17.7	1.34	0.0024	18.5	1.9	0.012
	256	26.4	0.84	< 0.001	26.1	1.52	0.004	27.2	1.32	0.053
	512	40.1	2.85	0.002	38.8	2.39	0.021	40.7	2.0	0.307

3.1   Results

The numerical results are presented in Table 2 (for Set1) and Table 3 (for Set2). Each table row presents the results averaged over 10 problem instances of the same type. For each algorithm (that is, A1, A2, and Beam-ACO) the results are provided in three columns. The first one (with heading **result**) provides the result of the corresponding algorithm averaged over 10 problem instances. The second column (with heading **std**) gives information on the corresponding standard deviation. Finally, the third column (with heading **time (s)**) provides the computation time. In the case of A1 and A2 the provided data corresponds to the total amount of computation time (averaged over 10 problem instances), while Beam-ACO was applied for 5 CPU seconds to each problem instance and column **time (s)** provides information on the time at which the best solution of a run was found (again, averaged over 10 problem instances).

Table 3. Experimental results concerning the instances of Set2.

$|\Sigma|$	#reps	Heuristic A1			Heuristic A2			Beam-ACO		
		result	std	time (s)	result	std	time (s)	result	std	time (s)
4	3	3.0	0.67	< 0.001	3.4	0.52	< 0.001	3.4	0.52	< 0.001
	4	3.4	0.52	< 0.001	3.8	0.42	< 0.001	3.8	0.42	< 0.001
	5	3.5	0.97	< 0.001	3.8	0.42	< 0.001	3.8	0.42	< 0.001
	6	3.6	0.52	< 0.001	3.8	0.42	< 0.001	3.8	0.42	< 0.001
	7	3.2	0.79	< 0.001	3.9	0.32	< 0.001	3.9	0.32	< 0.001
	8	3.8	0.42	< 0.001	4.0	0.0	< 0.001	4.0	0.0	< 0.001
8	3	5.5	0.85	< 0.001	5.8	0.63	< 0.001	5.9	0.57	< 0.001
	4	6.2	0.79	< 0.001	6.6	0.52	< 0.001	6.7	0.48	< 0.001
	5	6.1	1.10	< 0.001	6.9	0.88	< 0.001	6.8	0.79	< 0.001
	6	6.3	0.82	< 0.001	7.1	0.74	< 0.001	7.3	0.82	< 0.001
	7	6.8	0.63	< 0.001	7.5	0.53	< 0.001	7.6	0.52	< 0.001
	8	6.7	1.25	< 0.001	7.5	0.53	< 0.001	7.5	0.53	< 0.001
16	3	8.9	1.85	< 0.001	9.3	1.49	< 0.001	9.6	1.51	< 0.001
	4	9.9	1.37	< 0.001	10.6	1.07	< 0.001	11.1	1.1	0.0016
	5	12.0	1.63	< 0.001	12.3	1.16	< 0.001	13.7	1.25	0.21
	6	10.8	1.55	< 0.001	11.9	1.37	< 0.001	13.0	1.49	0.0044
	7	12.3	1.34	< 0.001	13.4	0.84	< 0.001	14.5	0.97	0.0072
	8	12.0	1.15	< 0.001	13.5	0.71	0.0012	14.7	0.95	0.038
32	3	14.9	1.97	< 0.001	14.7	1.34	0.0012	16.1	1.45	0.042
	4	16.9	1.91	< 0.001	17.1	1.20	< 0.001	19.2	1.55	0.013
	5	17.8	1.69	< 0.001	18.0	0.82	0.0016	20.6	0.84	0.103
	6	19.4	2.99	< 0.001	19.7	1.16	0.0016	24.0	2.11	0.45
	7	21.2	1.62	< 0.001	20.5	0.53	0.0024	24.9	1.37	0.048
	8	21.0	2.54	< 0.001	21.7	1.25	0.002	26.8	1.32	0.38
64	3	23.5	1.72	< 0.001	22.1	1.37	0.002	24.8	2.15	0.041
	4	27.7	2.16	< 0.001	25.2	1.40	0.0032	30.1	1.37	0.14
	5	30.4	2.12	< 0.001	27.0	1.25	0.0044	34.5	1.43	0.19
	6	33.4	2.22	< 0.001	29.1	1.20	0.0048	38.4	1.78	0.407
	7	36.9	3.45	0.0012	31.0	1.41	0.006	42.3	2.95	0.394
	8	37.1	3.14	0.0015	32.0	1.49	0.0084	45.1	2.23	0.916

First of all, regarding the computation times, it can be observed that all algorithms are very fast. In fact, even Beam-ACO usually finds the best solution of a run in a fraction of a second. Concerning the results, we can observe that Beam-ACO is nearly always superior (or equal) to both A1 and A2 on both instance sets. This is with the exception of one single case in instance set Set2 (alphabest size 8, and maximally 6 repetitions of the same letter) where heuristic A2 performs slightly better.

The graphics that are shown in Figs. 1 and 2 help to appreciate the improvement of Beam-ACO over A1 and A2. Hereby, Fig. 1 visualizes the improvement of Beam-ACO over A1, and Fig. 2 visualizes the (possibly negative) improvement of Beam-ACO over A2. The graphics in (a) (in both cases) concern Set1, whereas the graphics in (b) concern Set2. For each combination of sequence length and alphabet size (in the case of the graphics in (a)) and for each combination of the alphabet size and the maximal number of repetitions (in the case of the graphics in (b)) the size of the colored circle indicates the improvement of Beam-ACO over A1 (respectively A2) in percent. The legend links circle size with the scale of the percentages. Black circles indicate an improvement of Beam-ACO over A1

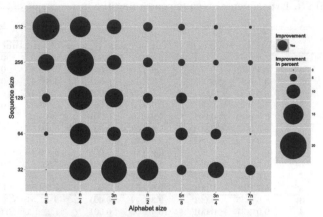

(a) Improvements (in percent) over A1 on instances of Set1.

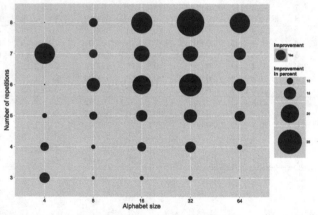

(b) Improvements (in percent) over A1 on instances of Set2.

Fig. 1. Percentage improvements of Beam-ACO over A1.

(respectively A2), while white circles indicate that the corresponding heuristic was better than Beam-ACO.

Concerning a comparison of Beam-ACO with A1 on instances of Set1, the graphic in Fig. 1(a) indicates that Beam-ACO has important advantages over A1 when the alphabet size is not too large. On the other side, when the alphabet size is rather large in comparison to the sequence length, the improvement of Beam-ACO over A1 decreases. Concerning Set2, the graphic in Fig. 1(b) indicates that Beam-ACO is generally much better than A1, with important advantages of up to $\approx 25\,\%$ improvement over A1 when the input sequences are long and the alphabet size is rather large. Concerning the comparison between Beam-ACO and A2, the graphics in Fig. 2 show that Beam-ACO clearly outperforms A2. For both instance sets Beam-ACO achieves up to 40–50 % of improvement over A2

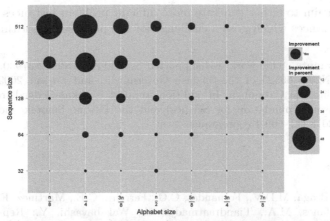

(a) Improvements (in percent) over A2 on instances of set 1.

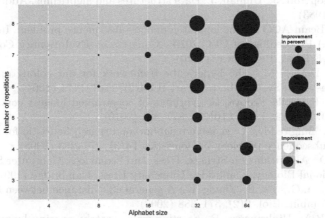

(b) Improvements (in percent) over A2 on instances of set 2.

Fig. 2. Percentage improvements of Beam-ACO over A2.

in the most difficult cases. Summarizing we can say that Beam-ACO is a new state-of-the-art method for the RFLCS problem.

4   Conclusions and Future Work

In this work we adapted a previously published Beam-ACO algorithm from the classical longest common subsequence problem to the repetition-free longest common subsequence problem. The results, in comparison to the best ones of the heuristics from the literature, show that Beam-ACO generally outperforms these heuristics, often even by a large margin.

Future work will include the development of a specific greedy function for the RFLCS problem, with the aim of using it within the Beam-ACO algorithm.

Moreover, we aim to either generate more difficult problem instances or to find real-world instances which pose a challenge for the proposed algorithm.

Acknowledgments. This work was supported by grants TIN2012-37930, TIN2010-14931 and TIN2007-66523 of the Spanish Government, and project 2009-SGR1137 of the Generalitat de Catalunya. In addition, support is acknowledged from IKER-BASQUE (Basque Foundation for Science) and the Basque Saiotek and Research Groups 2013-2018 (IT-609-13) programs.

References

1. Adi, S.S., Braga, M.D.V., Fernandes, C.G., Ferreira, C.E., Martinez, F.V., Sagot, M.F., Stefanes, M.A., Tjandraatmadja, C., Wakabayashi, Y.: Repetition-free longest common subsquence. Disc. Appl. Math. **158**, 1315–1324 (2010)
2. Aho, A., Hopcroft, J., Ullman, J.: Data structures and algorithms. Addison-Wesley, Reading (1983)
3. Blum, C.: Beam-ACO for the longest common subsequence problem. In: Fogel, G., et al. (eds.) Proceedings of CEC 2010 - Congress on Evolutionary Computation, vol. 2. IEEE Press, Piscataway (2010)
4. Blum, C., Dorigo, M.: The hyper-cube framework for ant colony optimization. IEEE Trans. Man Syst. Cybern. - Part B **34**(2), 1161–1172 (2004)
5. Bonizzoni, P., Della Vedova, G.: Variants of constrained longest common subsequence. Inf. Process. Lett. **110**(20), 877–881 (2010)
6. Easton, T., Singireddy, A.: A large neighborhood search heuristic for the longest common subsequence problem. J. Heuristics **14**(3), 271–283 (2008)
7. Gusfield, D.: Algorithms on Strings, Trees, and Sequences. Computer Science and Computational Biology, Cambridge University Press, Cambridge (1997)
8. Jiang, T., Lin, G., Ma, B., Zhang, K.: A general edit distance between RNA structures. J. Comput. Biol. **9**(2), 371–388 (2002)
9. Julstrom, B.A., Hinkemeyer, B.: Starting from scratch: growing longest common subsequences with evolution. In: Runarsson, T.P., Beyer, H.-G., Burke, E.K., Merelo-Guervós, J.J., Whitley, L.D., Yao, X. (eds.) PPSN 2006. LNCS, vol. 4193, pp. 930–938. Springer, Heidelberg (2006)
10. Maier, D.: The complexity of some problems on subsequences and supersequences. J. ACM **25**, 322–336 (1978)
11. Shyu, S.J., Tsai, C.Y.: Finding the longest common subsequence for multiple biological sequences by ant colony optimization. Comput. Oper. Res. **36**(1), 73–91 (2009)
12. Smith, T., Waterman, M.: Identification of common molecular subsequences. J. Mol. Biol. **147**(1), 195–197 (1981)
13. Storer, J.: Data Compression: Methods and Theory. Computer Science Press, Rockville (1988)

Applications

Medium-Voltage Distribution Network Expansion Planning with Gene-pool Optimal Mixing Evolutionary Algorithms

Hoang N. Luong[1(✉)], Marinus O.W. Grond[2], Peter A.N. Bosman[1], and Han La Poutré[1]

[1] Center for Mathematics and Computer Science (CWI), P.O. Box 94079, 1090 GB Amsterdam, The Netherlands
Hoang.Luong@cwi.nl
[2] Department of Electrical Engineering, Electrical Energy Systems, Eindhoven University of Technology (TU/e), P.O. Box 513, 5600 MB Eindhoven, The Netherlands

Abstract. Medium-voltage distribution network expansion planning involves finding the most economical adjustments of both the capacity and the topology of the network such that no operational constraints are violated and the expected loads, that the expansion is planned for, can be supplied. This paper tackles this important real-world problem using realistic yet computationally feasible models and, for the first time, using two instances of the recently proposed class of Gene-pool Optimal Mixing Evolutionary Algorithms (GOMEAs) that have previously been shown to be a highly efficient integration of local search and genetic recombination, but only on standard benchmark problems. One GOMEA instance that we use employs linkage learning and one instance assumes no dependencies among problem variables. We also conduct experiments with a widely used traditional Genetic Algorithm (GA). Our results show that the favorable performance of GOMEA instances over traditional GAs extends to the real-world problem at hand. Moreover, the use of linkage learning is shown to further increase the algorithm's effectiveness in converging toward optimal solutions.

Keywords: Evolutionary algorithms · Linkage learning · Distribution network · Power system expansion planning

1 Introduction

The Gene-pool Optimal Mixing Evolutionary Algorithm (GOMEA) combines genetic recombination as is reminiscent of Genetic Algorithms (GAs) with model-building as is reminiscent of Estimation of Distribution Algorithm (EDAs) and direct improvements as is reminiscent of Local Search (LS) [1]. The model used in GOMEA describes linkage relations between variables, i.e. which variables should be copied jointly when performing genetic recombination. Various sub-classes of the general linkage model are possible, ranging from allowing only fully

© Springer International Publishing Switzerland 2014
P. Legrand et al. (Eds.): EA 2013, LNCS 8752, pp. 93–105, 2014.
DOI: 10.1007/978-3-319-11683-9_8

independent linkage relations to allowing overlapping linkage relations. Based on the chosen and then learned linkage structure, GOMEA performs variation by intensively mixing building blocks as identified by the linkage relations in a greedy manner. The efficiency of GOMEA has so far been shown on a number of academic benchmarks [1,2], but not yet on real-world optimization problems.

A medium-voltage (MV) distribution network carries electricity from the (sub)transmission network to MV consuming units [3]. MV distribution network expansion planning (DNEP) is an important real-world engineering problem. As the loads (i.e. power consumptions) at different locations increase and/or newly appeared loads need connections to the network, various electrical components in the distribution network require replacement or new components must be installed. Both the capacities of the components and the topology of the network have to be taken into account. There exist various MV network layouts but the two most common topologies are: radial topology and open loop topology [3]. Radial topologies, in which every consuming unit is supplied by only one electrical feed path, are often used in distribution networks with overhead lines, especially for rural areas [3]. This paper focuses on the open loop layout, which is used for distribution networks with underground cables, typically found in urban areas of dense populations. Such MV networks contain groups of several consuming units (load points). In each group, consuming units are physically connected one by one by cables forming the shape of a loop. However, in normal operation, due to management and protection policies, one cable of every loop is put into an inactive state which creates an opening in the loop so that the network operates in a radial manner. Those cables are put in reserve to be used for reconfiguring the MV network when unexpected faults happen on active cables [3]. A feasible expansion plan is one that satisfies all operation and configuration constraints. An optimal plan is one that is feasible and has minimum expansion costs. In this paper, investment expenses are of sole interest.

There exist numerous studies into DNEP but the problem modelling is still far from being standardized. Every network operator has a different policy regarding the operation constraints of their power systems and different repositories of electrical facilities. Most studies evaluate the reliability of distribution networks based on the average failure rates and restoration times of components, in which reserve cables are considered as options to enhance the network reliability [4,5]. The result of such reliability analysis can then be capitalized into customer outage cost to include in the overall cost to be optimized [4] or can be treated as a separate objective function [5]. However, it is shown that reliability in practice is a relative index as its calculation involves many intricate problems with high uncertainty [6]. In this paper, we therefore consider the capacity of reserve cables, from a different and more practically relevant perspective, as a network configuration constraint, which is termed as reconfigurability. Reconfigurability requires the network to have enough reserve cables with adequate capacities to bring the network back to operation when an outage happens on some active cable. Although the cost function to be optimized is relatively simple and the problem variables are even pairwise independent in it, the constraint functions are far more involved and require dedicated electrical engineering computations (e.g. power flow calculations) that involve the entire network, effectively introducing dependencies between problem variables. It is therefore interesting, in addition to comparing the effectiveness of GOMEA with the commonly employed traditional GA, to see whether the use and usefulness of linkage learning also extends from traditional benchmark problems to real-world problems such as the one at hand.

The remainder of this paper is organized as follows. Section 2 outlines GOMEA and explains its components. Section 3 presents the anatomy of a conventional distribution network and important constraints. Section 4 shows and discusses the experimental results, while Sect. 5 concludes the paper.

2 Gene-pool Optimal Mixing Evolutionary Algorithm

Classic GAs have difficulty solving an optimization problem that has optimal solutions made of multivariate building blocks whose constitutive problem variables are scattered over the solution representation string [7]. Traditional recombination operators of GAs are either not able to juxtapose building blocks of nonconsecutive variables (i.e. 1- or 2-point, or uniform crossover) or too disruptive to preserve enough long building blocks (i.e. in case of uniform crossover). EDAs were developed with an emphasis on linkage learning to help to detect and preserve multivariate dependencies, but in EDAs this comes at the cost of estimating the complete probability distribution, which is expensive and may be unnecessary. On the other hand, problems with hierarchical dependencies provide a huge challenge for a classic GA as its genetic recombination is only horizontal and hierarchical dependencies (i.e. building blocks of building blocks) cannot be exploited directly. The reason for this is that there is no intermediate checking for improvements during genetic recombination, causing higher-level building blocks to automatically overwrite and undo the effects of mixing lower-level building blocks. GOMEA overcomes these issues by effectively integrating local search into variation, making its overall procedure closer to that of genetic local search [8]. For solving DNEP, GOMEA is therefore a strong candidate optimization algorithm.

2.1 Family of Subsets

The GOMEA uses the concept of family of subsets (FOS) as the linkage model to match the structure of optimization problems [1]. A FOS, denoted $\mathcal{F}$, is a set of subsets of a certain set S, which means $\mathcal{F} \subseteq \mathcal{P}(S)$, i.e. the powerset of S. Normally, set S is the set of all variable indices $\{1, 2, \ldots, l\}$. A FOS $\mathcal{F}$ can be written as $\mathcal{F} = \{\boldsymbol{F}^1, \boldsymbol{F}^2, \ldots, \boldsymbol{F}^{|\mathcal{F}|}\}$ where $\boldsymbol{F}^i \subseteq \{1, 2, \ldots, l\}$, $i \in \{1, 2, \ldots, |\mathcal{F}|\}$. To ensure all decision variables are considered in the variation operator, every variable index is contained in at least one subset in $\mathcal{F}$, i.e. $\forall i \in \{1, 2, \ldots, l\}$: $(\exists j \in \{1, 2, \ldots, |\mathcal{F}|\} : i \in \boldsymbol{F}^j)$. In this paper, we consider two FOS structures.

Univariate Structure: This structure, which is arguably the simplest structure possible, considers every decision variable to be independent from each other. The corresponding FOS $\mathcal{F}$ thus contains only singleton subsets $F^i = \{i\}, i \in \{1, 2, \ldots, l\}$. As there is only one possible configuration, no linkage learning is required. The use of the univariate structure is perhaps best known from GAs, where it translates into the well-known uniform crossover operator (UX).

Linkage Tree Structure: The linkage tree (LT) structure represents dependencies among decision variables in a hierarchical manner. The bottom level of the tree (i.e. leaf nodes) contains all singleton subsets, i.e. the univariate structure. Intermediate levels contains subsets F^i having more than one decision variable index. Any bivariate or multivariate subset F^i is the result of combining two subsets F^j and F^k such that $F^j \cap F^k = \emptyset, |F^j| < |F^i|, |F^k| < |F^i|$ and $F^j \cup F^k = F^i$. The top level (root node) is the set S itself containing all decision variable indices. This root node, which indicates that all variables are jointly dependent, is excluded from the linkage tree FOS as performing building block mixing based on this subset for any two solutions only results in the same solutions.

The LT is learned from the selected candidate solutions at every generation by performing a hierarchical clustering procedure where distances between clusters are computed using the average pair-wise distance over all pairs of variables. For details about clustering algorithms and different distance metrics, please refer to the literature [1,2]. Here, we used mutual information (MI) as the basis of distance between two variables (higher MI values mean a lower distance). We further note that in this paper, variables are not binary but rather have a larger bounded integer domain. However, since the search space is still Cartesian, the extension of MI from binary to integer variables is straightforward. The GOMEA variant that uses the LT structure as its linkage model is also known as Linkage Tree Genetic Algorithm (LTGA) [1]. It is worthwhile to mention that the computational complexity of learning an LT is low compared to typical higher-order models in EDAs (i.e. $O(nl^2)$ versus $O(nl^3)$).

2.2   Optimal Mixing and Forced Improvements

GOMEA uses a procedure called Gene-pool Optimal Mixing (GOM) as its variation operator [1]. For each existing parent solution in the population, exactly one offspring is generated by mixing building blocks of that parent with those of other solutions following the linkages specified by subsets in FOS $\mathcal{F}$. First, the parent solution is cloned. Then, the FOS is traversed and for each subset $F^i \in \mathcal{F}$ a donor solution is chosen randomly from the population. The values in the donor corresponding to the variables in the linkage group are copied into the parent solution. If such mixing results in an improvement, the changes are accepted, otherwise the changes are reverted. Bosman et al. [2] showed that if GOM also accepts changes that generate equally good solutions, better performance can be achieved.

If a solution cannot be improved by GOM alone, a procedure called forced improvement (FI) is performed [9]. In essence, FI is an additional GOM operation with the current best solution always as the donor. However, in this case, optimal mixing stops as soon as any single improvement is achieved. Because accepting solutions of equal quality can potentially stall the algorithm indefinitely on a fitness plateau, GOMEA is found to have better performance if FI is also triggered when the number of continuous generations that the best solution is not updated, which is termed as no-improvement stretch (NIS), is larger than $1 + \lfloor \log_{10}(n) \rfloor$ [2]. FI is reported to ensure efficient convergence while not continuously reducing population diversity [9]. The pseudo-code for GOMEA with GOM and FI is outlined in Fig. 1. Note that GOMEA typically does a lot more evaluations per generation than a classic GA would do, but GOMEA also typically requires far smaller population sizes and far less generations to converge.

GOMEA //*population size n*
  for $i \in \{1, 2, \ldots, n\}$ **do**
    $\mathcal{P}_i \leftarrow$ CREATERANDOMSOLUTION()
    EVALUATEFITNESS($\mathcal{P}_i$)
  $\boldsymbol{x}^{best} \leftarrow argmax_{\boldsymbol{x} \in \mathcal{P}}\{fitness[\boldsymbol{x}]\}$
  $t \leftarrow 0; t^{NIS} \leftarrow 0$
  while $\neg$TERMINATIONCONDITIONSSATISFIED **do**
    $\mathcal{S} \leftarrow$ TOURNAMENTSELECTION($\mathcal{P}, n, 2$)
    LEARNMODEL($\mathcal{S}$)
    for $i \in \{1, 2, \ldots, n\}$ **do**
      $\mathcal{O}_i \leftarrow$ FI-GOM($\mathcal{P}_i$)
    $\mathcal{P} \leftarrow \mathcal{O}$
    $\boldsymbol{x}^{best} \leftarrow argmax_{\boldsymbol{x} \in \mathcal{P}}\{fitness[\boldsymbol{x}]\}$
    if $fitness[\boldsymbol{x}^{best}(t)] > fitness[\boldsymbol{x}^{best}]$ **then**
      $t^{NIS} \leftarrow 0; \boldsymbol{x}^{best} \leftarrow \boldsymbol{x}^{best}(t)$
    else
      $t^{NIS} \leftarrow t^{NIS} + 1$
    $t \leftarrow t + 1$

FI-GOM($\boldsymbol{x}$)
  $b \leftarrow o \leftarrow \boldsymbol{x}; fitness[b] \leftarrow fitness[o] \leftarrow fitness[\boldsymbol{x}]; changed \leftarrow false$
  for $i \in \{1, 2, \ldots, |\mathcal{F}|\}$ **do**
    $p \leftarrow$ RANDOM($\{\mathcal{P}_1, \mathcal{P}_2, \ldots, \mathcal{P}_n\}$)
    $o_{\mathcal{F}i} \leftarrow p_{\mathcal{F}i}$
    if $o_{\mathcal{F}i} \neq b_{\mathcal{F}i}$ **then**
      EVALUATEFITNESS(o)
      if $fitness[o] \geq fitness[b]$ **then**
        $b_{\mathcal{F}i} \leftarrow o_{\mathcal{F}i}; fitness[b] \leftarrow fitness[o]; changed \leftarrow true$
      else
        $o_{\mathcal{F}i} \leftarrow b_{\mathcal{F}i}; fitness[o] \leftarrow fitness[b]$
  if $\neg changed$ or $t^{NIS} > 1 + \lfloor \log_{10}(n) \rfloor$ **then**
    $changed \leftarrow false$
    for $i \in \{1, 2, \ldots, |\mathcal{F}|\}$ **do**
      $o_{\mathcal{F}i} \leftarrow \boldsymbol{x}^{best}_{\mathcal{F}i}$
      if $o_{\mathcal{F}i} \neq b_{\mathcal{F}i}$ **then**
        EVALUATEFITNESS(o)
        if $fitness[o] > fitness[b]$ **then**
          $b_{\mathcal{F}i} \leftarrow o_{\mathcal{F}i}; fitness[b] \leftarrow fitness[o]; changed \leftarrow true$
        else
          $o_{\mathcal{F}i} \leftarrow b_{\mathcal{F}i}; fitness[o] \leftarrow fitness[b]$
      if $changed$ **then breakfor**
  if $\neg improved$ **then**
    $o \leftarrow \boldsymbol{x}^{best}; fitness[o] \leftarrow fitness[\boldsymbol{x}^{best}]$

Fig. 1. Pseudo-code for GOMEA [2]

3 MV Distribution Network Expansion Planning

Distribution network expansion planning (DNEP) involves decision making about what, where, when and how electrical components in a power distribution system should be adjusted to meet the forecasted growth in power demands at consuming units. In this paper, we take a traditional conservative approach and consider only the highest possible peak load for each consuming unit in the network. The network must be configured such that it can handle those loads and thus it is tested with that load profile. In this paper, we focus on a key part of the problem: deciding upon the locations and the types of adjustments. Available enhancement options are: changing existing devices, and installing new devices in the network, without specifying the time horizon. This paper considers two kinds of electrical devices: cables and transformers. An optimal expansion plan requires minimum investment cost while satisfying all operation and configuration constraints (see Sect. 3.2).

3.1 MV Distribution Network Encoding

An MV distribution network can be seen as a graph with a set of nodes (vertices) and a set of branches (edges). A node can be a substation, which is the source of power supply, or it can be a consuming unit, which demands and consumes power. Every branch connects two nodes, and all branches together form feed paths for electric currents flowing from power substations to consuming units. In a DNEP problem, the power supply capacities of substations and power demands of consuming units form the inputs. The outputs are decisions about capacities of all branches. Available options are: whether to connect two nodes by a branch (an overhead line or an underground cable, or a transformer if two nodes have different voltages), the capacity of the branch, and whether the branch should be active or in reserve.

To solve the DNEP for a network, we need to specify all the currently existing branches and a restricted set of potential candidate branches that can be newly added into the network. This set of potential branches is often determined by using expert knowledge to disregard unnecessary branches. Let l denote the total number of branches, and let m denote the total number of nodes. We represent a distribution network as a vector of length l of integer-value elements

$$\mathbf{x} = (x_1, x_2, \ldots, x_l), \quad x_i \in \Omega(x_i), \quad i \in \{1, 2, \ldots, l\} \tag{1}$$

where each x_i corresponds with the i-th branch of the network. The set of possible devices $\Omega(x_i)$ that can be installed at each branch x_i depends on policies and the repository of each network operator. We use an integer number to indicate which device to install at a branch. The status of each x_i is defined as follows

- $x_i = 0$: There is no device at the i-th branch. This means that the previously existing device is removed or that no device is decided to be installed at the i-th branch.

- $x_i = id > 0$: A device with identification number $id \in \Omega(x_i)$ is installed at the i-th branch.
- $x_i = -id < 0$: A device with identification number $id \in \Omega(x_i)$ is put in reserve at the i-th branch. The device is installed into the network but it does not take part in the normal operation. It is used to reconfigure the system in emergency cases.

Note that the original MV network has the x_i of currently non-existing branches set to 0.

3.2 Optimization Problem Formulation

Let $\mathbf{x} = (x_1, x_2, \ldots, x_l)$ be the original network and let $\mathbf{x}' = (x_1', x_2', \ldots, x_l')$ be an adjusted network. DNEP minimizes the investment cost as follows

$$\text{Min} \quad f(\mathbf{x}, \mathbf{x}') = \sum_{i=1}^{l} cost(x_i, x_i') \tag{2}$$

where

$$cost(x_i, x_i') = \begin{cases} 0 & \text{if } x_i = x_i' \\ \text{cost of changing } x_i \text{ to } x_i' & \text{if } x_i \neq x_i' \end{cases} \tag{3}$$

For a given (test) load profile, the following constraints must be satisfied:

I **Voltage constraints**

$$|V_i|^{MIN} \leq |V_i| \leq |V_i|^{MAX}, \quad i \in \{1, 2, \ldots, m\} \tag{4}$$

where $|V_i|$ is the voltage magnitude at node i, and $[|V_i|^{MIN}, |V_i|^{MAX}]$ is the allowable range of voltage magnitude at node i. We quantify the degree of the voltage constraint violation of a network by summing the amount of out-of-bound voltage magnitude at every node (i.e. $(|V_i|^{MIN} - |V_i|)$ if $|V_i| < |V_i|^{MIN}$ or $(|V_i| - |V_i|^{MAX})$ if $|V_i| > |V_i|^{MAX}$).

II **Line flow constraints (or device capacity constraints)**

$$|S_i| \leq |S_i|^{MAX}, \quad i \in \{1, 2, \ldots, l\} \tag{5}$$

where $|S_i|$ is the power flow through the device installed at branch x_i, i.e. a cable or a transformer, and $|S_i|^{MAX}$ is the nominal capacity of that device. There should be no overload at any device. We quantify the degree of the line flow constraint violation of a network by summing the amount of overload at every branch (i.e. $(|S_i| - |S_i|^{MAX})$ if $|S_i| > |S_i|^{MAX}$).

III **Radial operation constraint:** All the active cables together have to form a radial configuration. This means that any consuming unit is supplied electricity via one single feed path in normal operation.

IV **Reconfigurability constraint:** When, during normal operation, faults happen on an active branch, that branch is isolated from the network by opening its corresponding switches. The network is then reconfigured by closing the switches of reserve branches so that disconnected consuming units are served again. The network may operate with loops in an emergency situation and can endure a mild overload in a short time while the faulty branch is being repaired. The degrees of emergency capacity of equipments are decided by network operators. In this paper, we assume that equipment emergency capacity is 120 % of its nominal capacity.

Constraints I, II, and III are commonly adopted in the literature [4,5]. The constraint IV is employed here due to reasons mentioned in Sect. 1.

3.3   Solution Evaluation

As DNEP is a constrained optimization problem, the fitness evaluation for an expansion plan involves both the investment cost calculation and constraint evaluations. When we need to compare any two solutions, as in selection or the optimal mixing procedures, we use the concept of constraint domination. A feasible solution is one that satisfies all constraints. A feasible solution is always better than an infeasible one, a cheap feasible solution is better than a more expensive one, and if both solutions are infeasible then the one with less or equal degree of violation of all constraints and strictly less violation of at least one constraint is the better solution.

While calculating investment cost is a trivial operation, constraint evaluations are computationally expensive. For each expansion plan, we must perform a power flow calculation (PLC) [10] to obtain the value of the voltage at each node and the power flowing through each branch. These are used to check the constraints (I) and (II). In essence, a PLC involves solving a system of non-linear equations, called the AC power flow model. Due to inherent technical reasons, the commonly used cheaper linear DC model cannot be used for distribution network evaluation without a significant compromise on accuracy. For details of PLC, see e.g. [10]. Therefore, constraints evaluations are computationally expensive.

A complete fulfilment of the reconfigurability constraint requires performing a single-line contingency for every branch in the network: a branch is assumed to be failed, the network is then reconfigured back to operation, and the power flowing in each branch is re-calculated. This paper considers a computationally cheaper constraint evaluation commonly adopted in practice. It performs single line contingency only on cables branching directly from substations as these cables carry the heaviest loads before distributing power to subsequent nodes.

4   Experiments

4.1   Test Cases and Experiment Setup

Based on real-world data, we designed two MV distribution networks as optimization benchmarks.

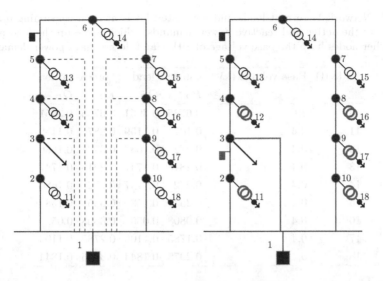

Fig. 2. Network 1. Original topology: Potential cables are represented by dashed lines. Reserve cables are marked with flag symbols. Transformers are denoted by pairs of overlapped circles. Arrow symbols indicate power demands at consuming units. After enhancement: Highlighted components are suggested to be replaced or newly installed.

- **Network 1:** an MV distribution network of one open loop contains 18 nodes (1 substation, 9 consuming units, in which each transformer is represented by 2 nodes having different base voltages) and 25 possible branches (10 existing cables, 8 existing transformers, and 7 potential cable connections). The topology and experiment current and forecasted loads of Network 1 can be found in Fig. 2 and Table 1.
- **Network 2:** an MV distribution network of two open loops contains 31 nodes (1 substation and 30 consuming units) and 59 possible branches (32 existing cables and 27 potential cable connections). Further details are withheld for reasons of confidentiality.

In this paper, we consider 5 common types of MV cables, differentiated by their areas of conductor: 120, 150, 240, 400, and 630 mm^2. We also consider 5 common options of transformers, denoted by their nominal capacities: 100, 160, 250, 400, and 630 kVA.

We test 3 optimizers: GOMEA-LT (GOMEA with linkage tree FOS), GOMEA-UNI (GOMEA with univariate FOS), and a traditional genetic algorithm (GA) with uniform crossover and tournament selection similarly configured as in [1]. For every optimizer, we test it with 10 different population sizes which are exponentially increased from 2^1 to 2^{10}. For every population size that we consider, we perform 30 independent runs of each optimizer. Each run starts with a population of randomly generated expansion plans (network topology and the equipment type at each element). We terminate a run only when the whole population converges

Table 1. Network 1: current loads and forecasted loads at each consuming unit. P_D and Q_D are the active and reactive power demands, which make up the load at each node. Other nodes have the base voltage of $10\,kV$ and do not have power demand.

Node ID	Base voltage (kV)	Current load		Forecasted load	
		P_D	Q_D	P_D	Q_D
3	10	0.6735	0.3951	3.6735	0.3951
11	0.4	0.187	0.1159	0.287	0.1159
12	0.4	0.272	0.1686	0.372	0.1686
13	0.4	0.2818	0.1747	0.2818	0.1747
14	0.4	0.272	0.1747	0.272	0.1686
15	0.4	0.255	0.158	0.355	0.158
16	0.4	0.0808	0.050	0.3808	0.05
17	0.4	0.1785	0.1106	0.2785	0.1106
18	0.4	0.2975	0.1844	0.3975	0.1844

to the same solution because in practice, the optimum is not know beforehand and we would like to see the best solutions that each optimizer possibly can obtain.

4.2   Results

Figure 2 shows MV Network 1 before enhancement and the best found expansion plan. To satisfy the forecast load demand, a new cable should connect node 1 (the substation) and node 3. The branch connecting node 2 and 3 should be put in reserve so that the network can operate radially. There are five overloaded transformers, and all of them should be replaced by ones with higher capacities.

Figure 3 shows the capability of GOMEA-LT, GOMEA-UNI, and GA in minimizing the investment cost for the enhancement of Network 1 as the number of fitness evaluations increases. Fitness evaluations for each candidate expansion plan involve power flow calculations, which are the most computationally expensive operations in the optimization process. Thus, different from academic benchmarks, fitness evaluation for the DNEP problem, truly dominates the computing time of all 3 optimizers. Hence, we use the number of fitness evaluations that each optimizer needs to perform from beginning until convergence as an indicator of computing time. Figure 3 shows both instances of GOMEA have better performances than the traditional GA. The traditional GA consumes much more computing time to come close to GOMEA but even for population size 1024, the traditional GA still cannot converge reliably to the same best solution obtained by GOMEA. If we use a too small population size, it is difficult to find feasible solutions, which explains why the line representing GA goes up first (feasible solutions can be more expensive than infeasible solutions) before it starts to go down when feasible solutions are found. Network 1 is a small distribution network containing only 25 branches (i.e. the number of decision variables), and

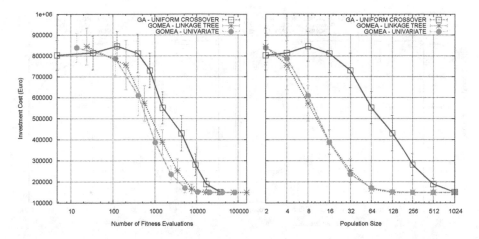

Fig. 3. Performance of GOMEA-LT, GOMEA-UNI and GA on minimizing the investment cost for enhancement of Network 1. Error bars show standard deviation.

while the variables are independent when evaluating the investment cost function, they are also linked when evaluating the constraints. However, depending on the problem instance, these linkages may be weak and of little influence, especially if the problem size is small. This explains why GOMEA-UNI, which assumes no dependencies among variables, requires less computing times than GOMEA-LT, which has an overhead of learning linkage trees and evaluating unnecessary mixings of (weak) linkage groups. This calls for the need of filtering spurious linkage groups in the linkage learning process as pointed out in [2]. It should be noted that when considering reliable convergence (30/30 runs) to the best solution ever found, GOMEA-LT requires less evaluations. The convenience of independent decision variables that GOMEA-UNI can exploit is not available in more complicated networks, which can be seen in case of Network 2.

Figure 4 shows the experimental results of 3 optimizers on solving DNEP for Network 2. This test case has a much larger and more complicated search space compared to Network 1. It can be seen that if we continue to run the optimization process with larger population sizes (and hence more power flow calculations), better solutions may still be obtained. Here, GOMEA-LT demonstrates that it has the best performance in comparison with the other 2 optimizers. The traditional GA has difficulty finding feasible solutions, let alone the optimum. GOMEA-UNI has a good performance here due to the intensive optimal mixing variation operator. However, without linkage learning, GOMEA-UNI does not obtain solutions of high quality as those found by GOMEA-LT. GOMEA-UNI can locate good solutions only if the decision variables are independent or weakly linked as in case of Network 1. Otherwise, GOMEA-UNI cannot efficiently find solutions that require the juxtaposition of multivariate linkage groups, e.g. as in the classic trap function benchmarks. GOMEA-LT wins over its univariate sibling in these cases.

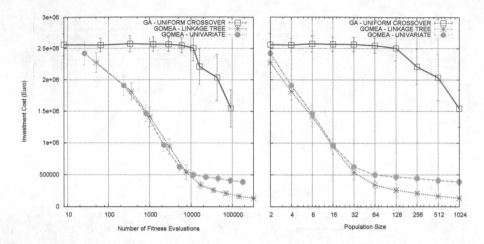

Fig. 4. Performance of GOMEA-LT, GOMEA-UNI and GA on minimizing the investment cost for enhancement of Network 2. Error bars show standard deviation.

5   Conclusions

The recently-developed gene-pool optimal mixing evolutionary algorithm (GOMEA) has so far been benchmarked on various theoretical optimization problems in the literature. Meanwhile, the long-existing traditional genetic algorithm (GA) has been widely used for numerous real-world optimization tasks. In this paper, we tackled the real-world problem of medium-voltage distribution network expansion planning (DNEP) with two instances of GOMEA: one with the univariate structure and one with the linkage tree. GOMEA was found to have much better performance than the traditional GA in terms of computing time and quality of the obtained solutions. Moreover, experimental results showed that linkage learning is truly beneficial for finding (near-)optimal solutions, not only in theoretical benchmarks but also in this engineering problem, further underlining the robustness of GOMEA and encouraging further applications of GOMEA on other real-world optimization problems.

References

1. Thierens, D., Bosman, P.A.N.: Optimal mixing evolutionary algorithms. In: Proceedings of the 13th Annual Genetic and Evolutionary Computation Conference, GECCO 2011, Dublin, Ireland, July 12–16, pp. 617–624. ACM (2011)
2. Bosman, P.A.N., Thierens, D.: More concise and robust linkage learning by filtering and combining linkage hierarchies. In: Genetic and Evolutionary Computation Conference, GECCO '13, Amsterdam, The Netherlands, July 6–10, pp. 359–366. ACM (2013)
3. Puret, C.: Mv public distribution networks throughout the world. Technical report 155, Merlin Gerin Group, March 1992

4. Falaghi, H., Singh, C., Haghifam, M.R., Ramezani, M.: Dg integrated multistage distribution system expansion planning. Int. J. Electr. Power Energ. Syst. **33**(8), 1489–1497 (2011)
5. Carrano, E.G., Soares, L.A.E., Takahashi, R.H., Saldanha, R.R., Neto, O.M.: Electric distribution network multiobjective design using a problem-specific genetic algorithm. IEEE Trans. Power Delivery **21**(2), 995–1005 (2006)
6. Slootweg, J.G., Van Oirsouw, P.M.: Incorporating reliability calculations in routine network planning: theory and practice. In: Proceedings of the 18th International Conference and Exhibition on Electricity Distribution - CIRED 2005, pp. 1–5 (2005)
7. Thierens, D., Goldberg, D.E.: Mixing in genetic algorithms. In: Proceedings of the 5th International Conference on Genetic Algorithms, Urbana-Champaign, IL, USA, pp. 38–47. Morgan Kaufmann, June 1993
8. Jaszkiewicz, A., Kominek, P.: Genetic local search with distance preserving recombination operator for a vehicle routing problem. Eur. J. Oper. Res. **151**(2), 352–364 (2003)
9. Bosman, P.A.N., Thierens, D.: Linkage neighbors, optimal mixing and forced improvements in genetic algorithms. In: Genetic and Evolutionary Computation Conference, GECCO '12, Philadelphia, PA, USA, July 7–11, pp. 585–592. ACM (2012)
10. Grainger, J.J., Stevenson, W.D.: Power System Analysis. McGraw-Hill Education, New York (2003)

Preliminary Studies on Biclustering of GWA: A Multiobjective Approach

Khedidja Seridi[1,2], Laetitia Jourdan[1,2](✉), and El-Ghazali Talbi[1,2]

[1] INRIA Lille - Nord Europe, DOLPHIN Project-Team,
59650 Villeneuve d'Ascq Cedex, France
[2] Université Lille 1, LIFL, UMR CNRS 8022, 59655 Villeneuve d'Ascq Cedex, France
{laetitia.jourdan,khedidja.seridi}@inria.fr, talbi@lifl.fr

Abstract. Genome-wide association (GWA) studies aim to identify genetic variations (polymorphisms) associated with diseases, and more generally, with traits. Commonly, a Single Nucleotide Polymorphism (SNP) is considered as it is the most common form of genetic variations. In the literature, several statistical and data mining methods have been applied to GWA data analysis. In this article, we present a preliminary study where we examine the possibilities of applying biclustering approaches to detect association between SNP markers and phenotype traits. Therefore, we propose a multiobjective model for biclustering problems in GWA context. Furthermore, we propose an adapted heuristic and metaheuristic to solve it. The performance of our algorithms are assessed using synthetic data sets.

1 Introduction

Association mapping has recently become a popular approach to discover the genetic causes of many complex diseases. A genome wide association study (GWAs) is the examination process of different genetic variants (markers) in several individuals in the purpose of detecting eventual association between the variants and certain traits. GWAs particularly focus on associations between single-nucleotide polymorphisms (SNPs) and traits like major diseases. Once such genetic associations are identified, researchers can use the information to promote new strategies to detect, treat and prevent the diseases [2].

Regarding the considered phenotype's nature, GWA studies usually deal with two classes of data. In the first class, the data comprise the genetic informations of all or a large fraction of the diseased subjects (cases) that appear in the considered study base and then sampling a comparable number of healthy subjects (controls), ideally from the same study base, and potentially matched with the cases by some socio-demographic characteristics such as race, age and gender. Accordingly, the considered trait is a qualitative trait *i.e.* an individual is even a case or a control. In the second class, the addressed phenotype is a quantitative trait *i.e.* numerical values that can be ordered from highest to lowest such as height, weight, cholesterol level, etc. The analysis of the later form of data is known as *Quantitative Trait Locus* (QTL) analysis.

© Springer International Publishing Switzerland 2014
P. Legrand et al. (Eds.): EA 2013, LNCS 8752, pp. 106–117, 2014.
DOI: 10.1007/978-3-319-11683-9_9

By considering the entire genome, case/control data analysis is essentially based on seeking alleles of variants that are more frequent in people with the disease (cases). The found variant is then said to be *associated* with the disease.

Quantitative trait locus (QTL) analysis is a statistical method that links two types of information *i.e.* phenotypic data (quantitative trait) and genotypic data (usually markers), in an attempt to explain the genetic basis of variation in complex traits [5]. QTL analysis allows researchers in different fields such as agriculture, evolution, and medicine to link certain complex phenotypes to specific regions of chromosomes. The goal of this process is to identify the action, interaction, number, and precise location of these regions.

A QTL analysis starts by collecting phenotype and genotype data from a number of unrelated individuals in the same way as in a case-control study. However, in QTL studies there are no cases and no controls, just individuals with a range of phenotype values. After that, association between the traits and the different SNPs are detected using statistical method. The associations are commonly formulated as predictive models.

Generally, genome wide associations studies are performed using supervised methods such as logistic regression and discriminant analysis [1,9], Bayesian approaches [4], etc. Commonly, the treated data comprises two main informations for each individual: genotype informations and phenotype informations. Using a training data set, the study mainly consists in defining a predictive model and validate it through a test data set.

In this work we propose an unsupervised study of the GWA data with quantitative traits (QTL). By this study we aim to extract a subset of SNPs that have the same alleles for a sub set of individuals sharing similar traits. Actually, the considered data can be seen as a matrix $A = (X, (Y, Z)) = \{a_{ij}\}$ where each row i presents an individual, each column j represents either a SNP ($j \in Y$) or a trait ($j \in Z$) and an element a_{ij} presents the corresponding SNP's allele (if $j \in Y$) or the corresponding traits value (if $j \in Z$) (see Table 1). Thus, a bicluster $B = (I, (J, K))$ is a sub-matrix of $A = (X, (Y, Z))$ where $I \subset X, J \subset Y$ and $K \subset Z$.

This paper is organized as follows. Section 2 presented the biclustering problem and a new multiobjective model for a biclustering problem applied to analyzing GWA data sets. An adapted heuristic and metaheuristic are proposed in

Table 1. Studied GWA data

	SNPs			Traits		
	S_1	...	S_A	T_1	...	T_B
A_1	a_{11}	...	a_{1A}	a_{1A+1}	...	a_{1M}
...	...	...	...	...	...	...
A_i	a_{i1}	...	a_{iA}	a_{iA+1}	....	a_{iM}
...	...	...	...	...	...	...
A_N	a_{N1}	...	a_{NA}	a_{NA+1}	....	a_{NM}

Sect. 3 to solve the proposed model. In Sect. 4, experimental analysis of the proposed approaches and results are presented. Finally Sect. 5 concludes the paper and presents perspectives.

2   Biclustering Method in Analyzing GWA Data

2.1   Biclustering

Biclustering or co-clustering is a well-known data mining method that has been widely applied in a broad range of domains such as marketing, psychology and bioinformatics. It consists in extracting submatrices $B = (I, J)$ $(I \subset X, J \subset Y)$ (called biclusters) with maximal size and respecting a certain coherence constraint. Depending on the addressed problem, biclusters of different types can be considered. The different biclusters types and some corresponding applications are described below.

1. Constant bicluster: all the biclusters elements have the same value.
2. Bicluster with constant rows/columns: the elements of each row (column) have the same value.
3. Bicluster with coherent values: the definition of this type of biclusters is a generalization of constant rows/columns biclusters. There exist two different models associated to this class of biclusters:
   (a) shifting model: where each row (and each column) can be obtained by adding an offset to an other row (column).
   (b) scaling model: where each row (and each column) can be obtained by multiplying an other row (column) by a factor.
4. Bicluster with coherent evolution: the elements of the bicluster behave similarly (correlated) independently of their numerical values.

When formulating a biclustering problem, a similarity (dissimilarity) measure is required in order to evaluate the extracted results. The measure is, commonly, related to the bicluster's type. In the case of microarray data analysis, the study aim to extract biclusters with coherent values or evolution (gene that present similar behavior under a sub set of conditions). Different multiobjective modeling for biclustering problem for microarrays data have been proposed [7,10–14] but none for the case of GWA data. Commonly, the proposed multiobjective models comprise: one or more function(s) to optimize the biclusters sizes, a function that optimizes biclusters coherences and a function to optimize the rows variances. In all of these models, a solution represents one bicluster. Regarding the size, most of the models maximize the ratio between the biclusters elements number and the microarray data elements. However, as the number of rows is generally more important than the number of columns, such functions may favor the maximization of rows number with regard to columns number. Thereby, in [7], authors proposed to maximize the number of rows and columns separately by using two objective functions. Concerning biclusters coherence, all the proposed models consider the Mean Squared Residue MSR [3] dissimilarity measure.

In [14] the MSR value is allowed to increase as it does not exceed the threshold δ. Regarding the rows fluctuations, all the existing models maximize the mean row variance. In [12] the coherence and fluctuation objectives are merged in one function by defining a function as the ratio between the MSR of the bicluster and its mean rows variance.

The MSR measure is well adapted to identify biclusters with coherent values. However, this measure can not be applied for GWA data as different biclusters type is required.

2.2   Multiobjective Problem Modeling

In this section, we propose a multiobjective model for a biclustering method applied to GWAs. In this study, we seek to extract biclusters with constant columns, which correspond to a set of individuals that share SNPs presenting the same alleles and the same traits. In order to extract such biclusters, two objectives have to be considered: maximizing the biclusters size (find maximal biclusters) and minimizing the average of columns variances. Actually, these two criteria are clearly independent and conflicting. In fact, a non perfect bicluster's coherence (columns constance) can be improved by removing a row or a column, *i.e.* by reducing its size. We can therefore deduce that the problem of biclustering in GWAs can be formulated as a multiobjective optimization problem. Thus, the proposed model is given by:

$$f_1(I,(J,K)) = \alpha \times \tfrac{|I|}{|X|} + \beta \times \tfrac{|J|}{|Y|} + \gamma \times \tfrac{|K|}{|Z|}$$

$$f_2(I,(J,K)) = Avar(I,(J,K)) = \tfrac{1}{|I| \times (|J|+|K|)} \sum_{j \in J \cup K} \sum_{i \in I} (a_{ij} - a_{Ij})^2$$

Where f_1 (size) has to be maximized and f_2 (average variance) has to be minimized

3   Resolution Approaches

In this section we present two new approaches to solve the proposed model. The first approach is a greedy heuristic *Sbic* and the second approach is a multiobjective metaheuristic $SHMOBI_{ibea}$.

3.1   Sbic Heuristic

Sbic is a greedy heuristic that aims to extract relevant biclusters from GWA data matrix and that has been designed in a similar manner as Cheng and Churchs heuristic [3] widely used for microarray data. At each run, *Sbic* extracts one bicluster from the data matrix. *Sbic* deletes (adds) nodes that meet with some conditions in order to decrease the biclusters average columns variances and increase its size. The main steps of *Sbic* are given in Algorithm 1.

In multiple node deletion phase, *Sbic* starts by removing some nodes (rows and columns) in order to decrease the average columns variance. In columns

---

Algorithm 1. Sbic Algorithm

1:**Input:** Bicluster $(I, (J, K))$ /*which can be the whole data matrix*/
2: **if**$(Avar(I, (J, K)) > \delta)$
3:     MultipleNodeDeletion(I,J,K)
4:     **if**$(Avar(I, (J, K)) > \delta)$
5:         SingleNodeDeletion (I,J,K)
6:     **endif**
7: **endif**
8: MultipleNodeAddition(I,J,K)

---

dimension, the variance of each column is calculated. The columns that have the highest variance are deleted. This process will clearly decrease the whole average variances of the columns. Similarly, the average variance can also be decreased by applying the same process on the rows dimension. Indeed, rows with the highest contribution on the average columns variances are deleted. After that, if the bicluster's average variance still higher than δ the bicluster has to undergo the single node deletion processes. The main steps are illustrated in Algorithm 2.

Algorithm 2. Multiple node deletion

1:**Input:** Bicluster $(I, (J, K))$
2: Compute a_{Ij}, $Avar$ and $con_i = \frac{\sum_{j \in J}(a_{Ij} - a_{ij})^2 + \sum_{k \in K}(a_{Ik} - a_{ik})^2}{|J| + |K|}$ $i \in I$
3: **if**$(con_i > \gamma \times Avar)$
4:     Remove the rows $i \in I$
5: **endif**
6: Compute a_{Ij}, $Avar$ and var_j $j \in J$
7: **if**$(var_j > \gamma \times Avar)$
8   Remove the column $j \in J$
9: **endif**
10: Compute a_{Ik}, $Avar$ and var_k $k \in K$
11: **if**$(var_k > \gamma \times Avar)$
12:     Remove the column $k \in K$
13: **endif**

In single node deletion, the nodes with the highest contribution on the average variance are iteratively deleted until the $Avar$ reaches the desired value. The main steps are illustrated in Algorithm 3.

Once the $Avar$ of the considered bicluster reaches the desired value, the algorithm tries to add other rows (columns) without increasing the $Avar$. For instance all the columns (not present yet in the bicluster) that have a variance lower than or equal to $Avar$ are added to the bicluster. Furthermore, the expected contribution of each row i (con_i) in the biclusters $Avar$ value is computed in order to decide whether the row can be added to the bicluster or not.

Algorithm 3. Single node deletion

1:**Input:** Bicluster $(I, (J, K))$
2:     **while**$(Avar(I, (J, K)) > \delta)$
3:       Recompute con_i, var_j and var_k.
4:       Find the node d (row or column) with the highest var_d (con_d) .
5:       Delete d.
6:     **endwhile**

The main steps are illustrated in Algorithm 4.

Algorithm 4. Multiple node addition

1:**Input:** Bicluster $(I, (J, K))$
2: Compute a_{Ij}, $Avar$ and $con_i = \frac{\sum_{j \in J}(a_{Ij} - a_{ij})^2 + \sum_{k \in K}(a_{Ik} - a_{ik})^2}{|J| + |K|}$ $i \notin I$
3: **if**$(con_i \leq Avar)$
4:     Add the rows i
5: **endif**
6: Compute a_{Ij}, $Avar$ and var_j $j \notin J$
7: **if**$(var_j \leq Avar)$
8:     Add the column j
9: **endif**
10: Compute a_{Ik}, $Avar$ and var_k $k \notin K$
11: **if**$(var_k \leq Avar)$
12:     Add the trait k
13: **endif**

Actually, *Sbic* is a deterministic algorithm. Thus, the same bicluster will be extracted if the starting matrix is always the same. In order to extract several biclusters from a data matrix $(X, (Y, Z))$ we propose to apply the *Sbic* over the whole data matrix to extract the first bicluster. After that, *Sbic* can be applied over a sub-matrix containing $p\%$ of the data's rows and columns selected randomly which will lead to discovering different bicluster at each run.

In the following section we present the main components of $SHMOBI_{ibea}$ metaheuristic.

3.2 $SHMOBI_{ibea}$

$SHMOBI_{ibea}$ is based on $HMOBI_{ibea}$ [15] which is a multiobjective meta-heuristic based on the evolutionary algorithm $MOBI_{ibea}$ [6] and $DMLS$ $(1 \cdot 1_{\succ})$ [8].

MOBI is a hybrid MOEA (Multi Objective Evolutionary Algorithm) for solving biclustering problem in the specific case of microarray data. It combines

IBEA with a local search inspired from Cheng and Churchs heuristic [3] which is dedicated for biclustering of microarray data. $MOBI_{ibea}$ [6] allows in the case of microarray data to extract biclusters of good quality.

DMLS (Dominance-based Multiobjective Local Search) are a general concept of multiobjective local searches using the concept of Pareto Optimality. At each generation, DMLS selects one or more non-visited solutions (solutions with non-explored neighborhood) from the *archive* and explores their neighborhoods. After that, the solutions are marked as visited. Different variants of DMLS exists depending on the number of selected solutions and on the exploration strategy. In this study, we will use $DMLS(1 \cdot 1_{\succ})$ where one solution is randomly selected and the exploration of its neighborhood stops when the first improving solution is found.

In this section, we propose $SHMOBI_{ibea}$ which is an adapted version of $HMOBI_{ibea}$ to SNP data. Several changes have been done to adapt $HMOBI_{ibea}$ to the specific case of SNPs. Therefore, we present a suitable solutions encoding and variation operators.

Solutions Encoding. In $SHMOBI_{ibea}$, we choose to represent a bicluster as a list compound of six parts: Each one of the first 3 parts of the chromosome is an ordered list of indexes corresponding to either rows, columns or traits; while parts 4 to 6 are just the cardinalities of those lists.

Example:
Given the data matrix presented in Table 2, the string $\{1\ 3\ 2\ 3\ 2\ 2\ 2\ 1\}$ represents the following bicluster compound of two rows (1 and 3), two SNPs (2 and 3) and one trait (2):

$$\{1\ 3\ 2\ 3\ 2\ 2\ 2\ 1\} \implies \begin{bmatrix} 2\ 1 & 0.3 \\ 0\ 0 & -0.75 \end{bmatrix}$$

Variation Operators.

1. **Crossover**:
   A Single point crossover is used in the three first parts of the solution (rows part, columns part and traits part). Each part undergoes crossover separately. Let parents be chromosomes $P_1 = \{r_1 \ ... \ r_n \ c_1 \ ... \ c_m \ t_1 \ ...t_p \ r_{nb} \ c_{nb} \ t_{nb}\}$ and $P_2 = \{r'_1 \ ... \ r'_l \ c'_1 \ ... \ c'_k \ t'_1 \ ... \ t'_q \ r'_{nb} \ c'_{nb} \ t'_{nb}\}$ where $r_n \leqslant r'_l$.

Table 2. Example of SNPs and traits data matrix

SNPs			Traits	
1	**2**	1	12.5	**0.3**
0	1	2	10.75	1.2
1	**0**	**0**	10.33	**−0.75**

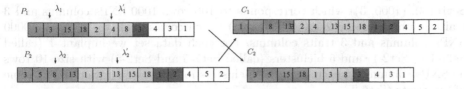

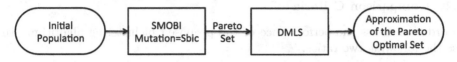

Fig. 1. An example of the crossover operator application.

Fig. 2. General scheme of $SHMOBI$

The crossover in the rows part is performed as follows: The crossover point in P_1 (λ_1) is generated as a random integer in the range $2 \leqslant \lambda_1 \leqslant r_n$. the crossover point in P_2 $\lambda_2 = r'_j$ where $r'_j \geqslant \lambda_1$ and $r'_{j-1} \leqslant \lambda_1$. In the same way, the crossover in the columns part and traits part is performed. The parts 4–6 are not involved directly in the crossover and are computed after it.

For example, consider the parents P_1 and P_2 presented in Fig. 1. Suppose the 3rd gene index and the 2nd condition index of P_1 are selected, so: $\lambda_1 - 15$ and $\lambda'_1 = 5$ then $\lambda_2 = 16$ and $\lambda'_2 = 6$, which results on the offspring C_1 and C_2.

2. **Mutation:**
   We replace the mutation operator by the *Sbic* heuristic.

   When generating random biclusters, it may happen that irrelevant rows and columns get included in spite of their values lying far apart. Therefore, we start by randomly generating a population where the irrelevant rows and columns of each bicluster are deleted using the *Sbic* heuristic. The resulting population is used as the initial population for $SMOBI_{ibea}$. After that, the $DMLS(1 \cdot 1_{\succ})$ is applied for each solution of $SMOBI_{ibea}$'s archive (Pareto approximation). The main steps of $SHMOBI_{ibea}$ are illustrated in Fig. 2.

4   Experiments and Results

In this section we present the experimental protocol in assessing the performance of the presented algorithms over synthetic data sets.

4.1   Data Sets

In order to assess the performance of the proposed algorithms, we use synthetic data sets to investigate the ability of our algorithms to extract implanted biclusters. In this purpose, we randomly generate different data sets of size:

Set1(100, (1000, 3)) which corresponds to 100 rows 1000 SNPs columns and 3 traits columns and Set2(100, (10000, 3)) which corresponds to 100 rows 10000 SNPs columns and 3 traits columns. For each data set we implant 1 (called Set1-1 et Set2-1) and 5 biclusters (called Set1-5 and Set2-5) with size 10 rows 50 SNPs columns. In each case, the biclusters may involve all (Set*-A) or some of the traits (Set*-T).

4.2   Comparison Criteria

In order to assess the performance of the proposed biclustering algorithm, we use the following two ratios:

$$\theta_{Shared} = \frac{S_{cb}}{Tot_{size}} \times 100 \tag{1}$$

Where S_{cb} is the portion size of bicluster correctly extracted and Tot_{size} is the total size of the implanted bicluster.

$$\theta_{NotShared} = \frac{S_{ncb}}{Tot'_{size}} \times 100 \tag{2}$$

Where S_{ncb} is the portion size of bicluster not correctly extracted and Tot'_{size} is the total size of the extracted bicluster.

The ratio θ_{Shared} (resp. $\theta_{NotShared}$) expresses the rate of shared (resp. not shared) biclusters volume with real biclusters. In fact, when θ_{Shared} (resp. $\theta_{NotShared}$) is equal to 100 % the algorithm extracts the correct (resp. not correct) biclusters. A perfect solution has θ_{Shared} =100 % and $\theta_{NotShared}$=0 % respectively. That is, the exact number of rows and columns of implanted biclusters.

4.3   Parameters

Concerning the models parameters, we set α, β and γ to 0.5, 0, 0.5 respectively. In fact, given the nature of data, SNPs columns present low variance compared to trait columns. Hence, a big number of SNP columns will be added for each bicluster undergoing the *Sbic* heuristic. Therefore, we favor biclusters having low average variance and low SNPs columns to be selected in the search process and this by setting $\beta = 0$.

In the other hand, all algorithms parameters have been set experimentally. For the *Sbic* we set α to 1.5, δ to 0.15 and %p to 50 %. The algorithm is run 20 times in order to extract 20 biclusters. The first run uses all the data matrix. The remaining runs starts by sub-matrices where the rows and the columns are chosen randomly. When selecting rows, more chance is given to rows not present yet in the previously extracted biclusters.

Concerning $SMOBI_{ibea}$, we experimentally set the initial population size to 400. The mutation and crossover operators parameters are set to 0.2 and 0.5

respectively. The algorithm stops after a fixed time depending on the data set size. For Set1 data sets the execution time is set to 500 s, and 700 s for Set2 data sets. The same time is allocated to $SHMOBI_{ibea}$ algorithm where 90 % of the execution time is accorded to $SMOBI_{ibea}$ and the remaining 10 % to $DMLS(1 \cdot 1_{\succ})$.

We apply our algorithms on the considered data sets and for each algorithm we select the closest biclusters to the implanted ones. Thereafter, we calculate θ_{Shared} and $\theta_{NotShared}$ for each bicluster. For instances where several biclusters are implanted, we report the average θ_{Shared} and $\theta_{NotShared}$ of the extracted biclusters.

4.4 Results

In this section, we compare the efficiency of $Sbic$, $SMOBI_{ibea}$ and $SHMOBI_{ibea}$ in extracting the implanted biclusters. The comparison is done with regard to θ_{Shared}, $\theta_{NotShared}$ and the rate of found biclusters.

Tables 3 and 4 present the obtained results for the different instances corresponding to one and five implanted biclusters respectively. A detailed observation of the found solutions show that, in most cases, the not correctly biclusters extracted portions are mainly composed of extra columns (SNPs).

In Table 3 we can observe that all the approaches can find the implanted bicluster. However, $SHMOBI_{ibea}$ find the best results with the highest θ_{Shared} and lowest $\theta_{notShared}$. For instance, in the case of data Set1-1-A where all the traits are involved in the bicluster, $SHMOBI_{ibea}$ extracts the bicluster with only $\theta_{notShared} = 24.24$ %. Actually, $SMOBI_{ibea}$ is able to find the implanted bicluster. However, the $\theta_{NotShared}$ of the extracted bicluster is very high. This result demonstrates the role of $DMLS(1, 1_{\succ})$ in fine-tuning the found results.

Table 3. Comparative results when extracting one bicluster. $SMOBI$ stands for $SMOBI_{ibea}$, $SHMOBI$ for $SHMOBI_{ibea}$.

Data	θ_{Shared}			$\theta_{NotShared}$			Rate of found biclusters		
	Sbic	SMOBI	SHMOBI	Sbic	SMOBI	SHMOBI	Sbic	SMOBI	SHMOBI
Set1-1-A	78.6 %	100 %	100 %	86.6 %	80.07 %	24.24 %	100 %	100 %	100 %
Set2-1-A	100 %	100 %	100 %	86.07 %	86.73 %	57.01 %	100 %	100 %	100 %
Set1-1-T	60.0 %	100 %	100 %	78.05 %	92.46 %	76.36 %	100 %	100 %	100 %
Set2-1-T	30 %	90 %	100 %	81.41 %	95.12 %	67.12 %	100 %	100 %	100 %

Table 4. Comparative results when extracting five biclusters. $SMOBI$ stands for $SMOBI_{ibea}$, $SHMOBI$ for $SHMOBI_{ibea}$.

Data	θ_{Shared}			$\theta_{NotShared}$			Rate of found biclusters		
	Sbic	SMOBI	SHMOBI	Sbic	SMOBI	SHMOBI	Sbic	SMOBI	SHMOBI
Set1-5-A	50.62 %	52.5 %	85.92 %	86.37 %	85.61 %	60.9 %	60 %	80 %	80 %
Set2-5-A	63.33 %	64.16 %	85.83 %	92.87 %	91.11 %	92.26 %	60 %	60 %	60 %
Set1-5-T	41.66 %	64.88 %	62.61 %	87.27 %	82.04 %	82.44 %	40 %	80 %	80 %
Set2-5-T	45 %	75 %	85.83 %	98.21 %	94.5 %	93.47 %	40 %	40 %	60 %

Similarly, Table 4 shows that $SHMOBI_{ibea}$ outperforms $Sbic$ and $SMOBI_{ibea}$ in finding the implanted biclusters. Actually, $SHMOBI_{ibea}$ finds more biclusters than the other approaches with higher θ_{Shared}. However, the $\theta_{NotShared}$ value of the biclusters extracted using all the approaches are relatively high. This can be explained by the huge number of SNPs columns in the data set.

Concerning running times, they are of 500 s for small instances (Set1-*) and 700 s for large instances (Set2-*).

5 Conclusion

In this article, we have presented a preliminary study on using a biclustering method to analyze GWA data. Actually, GWA data consists in two types of information i.e. phenotype data (traits) and genotype data (genetic variations). Commonly, SNPs are considered as they present the most frequent form of genetic variations. The analysis of such data consists in finding eventual associations between traits and SNPs combinations. Therefore, we propose a multiobjective modeling for biclustering in order to extract samples (individuals) sharing similar traits and having same alleles for a SNPs combination. The corresponding biclusters are constant columns biclusters.

The extracted biclusters may bring out existing associations between the considered SNPs and traits. Moreover, the extracted biclusters may provide important informations that can be used in further GWA studies. Given the huge number of SNPs, we propose to solve this problem using a hybrid metaheuristic $SHMOBI_{ibea}$. The efficiency of $SHMOBI_{ibea}$ have been assessed using synthetic data sets of different sizes and different implanted biclusters numbers. Further studies will be carried out in real data sets provided by the company *Genes Diffusions*[1].

References

1. Binder, H., Tina, M., Holger, S., Klaus, G., Michael, S., Jan, H., Katja, I., Martin, S.: Cluster-localized sparse logistic regression for SNP data. Stat. Appl. Genet. Mol. Biol. **11**(4), 13 (2012)
2. Bush, W.S., Moore, J.H.: Chapter 11: Genome-wide association studies. PLoS Comput. Biol. **8**(2), e1002822 (2012)
3. Cheng, Y., Church, G.M.: Biclustering of expression data. In: Proceedings of the 8th ISMB, pp. 93–103. AAAI Press, Menlo Park (2000)
4. Ding, L., Baye, T.M., He, H., Zhang, X., Kurowski, B.G., Martin, L.J.: Detection of associations with rare and common SNPs for quantitative traits: a nonparametric Bayes-based approach. BMC Proc. **5**(Suppl 9), S10 (2011)
5. Douglas, F., Trudy, M.: Introduction to Quantitative Genetics, 4th edn. Prentice Hall, Englewood Cliffs (1996)
6. Seridi, K., Jourdan, L., Talbi, E.-G.: Multi-objective evolutionary algorithm for biclustering in microarrays data. In: IEEE Congress on Evolutionary Computation, pp. 2593–2599. IEEE (2011)

[1] http://www.genesdiffusion.com/default.aspx

7. Lashkargir, M., Monadjemi, S.A., Dastjerdi, A.B.: A new biclustering method for gene expersion data based on adaptive multi objective particle swarm optimization. In: Proceedings of the 2009 Second International Conference on Computer and Electrical Engineering. ICCEE '09, vol. 01, pp. 559–563. IEEE Computer Society, Washington, DC, USA (2009)

8. Liefooghe, A., Humeau, J., Mesmoudi, S., Jourdan, L., Talbi, E.-G.: On dominance-based multiobjective local search: design, implementation and experimental analysis on scheduling and traveling salesman problems. J. Heuristics **18**(2), 317–352 (2012)

9. Lin, H., Desmond, R., Bridges, S.L., Soong, S.: Variable selection in logistic regression for detecting SNP-SNP interactions: the rheumatoid arthritis example. Eur. J. Hum. Genet. **16**(6), 735 (2008)

10. Liu, J., Chen, Y.: Dynamic biclustering of microarray data with MOPSO. In: 2010 IEEE International Conference on Granular Computing. GrC 2010, pp. 330–334. IEEE Computer Society, San Jose, CA, USA, 14–16 August 2010

11. Liu, J., Li, Z., Hu, X., Chen, Y.: Biclustering of microarray data with MOSPO based on crowding distance. BMC Bioinf. **10**(Suppl 4), S9 (2009)

12. Liu, J., Li, Z., Hu, X., Chen, Y.: Multi-objective ant colony optimization biclustering of microarray data. In: Granular Computing, pp. 424–429 (2009)

13. Liu, J., Li, Z., Liu, F., Chen, Y.: Multi-objective particle swarm optimization biclustering of microarray data. In: IEEE International Conference on Bioinformatics and Biomedicine, pp. 363–366 (2008)

14. Mitra, S., Banka, H.: Multi-objective evolutionary biclustering of gene expression data. Pattern Recogn. **39**(12), 2464–2477 (2006)

15. Seridi, K., Jourdan, L., Talbi, E.-G.: Hybrid metaheuristic for multi-objective biclustering in microarray data. In: CIBCB, pp. 222–228. IEEE (2012)

An Evolutionary Approach to Contrast Compensation for Dichromat Users

A. Mereuta$^{(\boxtimes)}$, S. Aupetit, N. Monmarché, and M. Slimane

Laboratoire Informatique (EA6300), Université François Rabelais Tours,
64, avenue Jean Portalis, 37200 Tours, France
{alina.mereuta,aupetit,monmarche,slimane}@univ-tours.fr

Abstract. In this paper, we are focusing on web accessibility, more precisely on improving web accessibility for Color Vision Deficiency (CVD) users. The contrast optimization problem for dichromat users can be modeled as a mono objective function which at minimum provides a suitable solution to the problem. The function aims to compensate the loss and maintains simultaneously a minimum change in the original color. The CMA-ES method is used to minimize the function. Experiments were conducted on real and artificial data in order to assess the approach efficiency for different set of parameters. The results showed that it is likely that the method performs better when the loss is important. The approach produces satisfying results on both real and artificial data for the set of tested parameters.

Keywords: Assistive technologies · Color vision deficiency · Dichromacy · CMA-ES

1 Web Accessibility and CVD Users

Web accessibility translates through full access to web resources for all users. Sets of recommendations were proposed by the W3C's WAI (World Wide Consortium' Web Accessibility Initiative) [17]. Web Content Accessibility Guidelines (WCAG) 1.0 [16] proposed in May 1999 provides recommendations for creating accessible web content. They are organized as a set of general principles. For each principle, conformance levels are defined according to their impact on web accessibility and a set of checkpoints is defined. The newest version of the guidelines (WCAG 2.0) regroups the recommendation into 4 main categories: perceivable, operable, understandable and robust. For each category, a list of success criteria is presented. For each criterion, several techniques are presented to achieve it. Even though efforts were made by webmasters, very little interest on accessibility is shown while building and designing web sites. In this work, we are focusing on color contrast improvement for users with color vision deficiency, more precisely dichromat users. WCAG 1.0 states that "Ensure that foreground and background color combinations provide sufficient contrast when viewed by someone having color deficits or when viewed on a black and white screen".

© Springer International Publishing Switzerland 2014
P. Legrand et al. (Eds.): EA 2013, LNCS 8752, pp. 118–128, 2014.
DOI: 10.1007/978-3-319-11683-9_10

Color vision deficiency (CVD) is the inability to perceive correctly certain colors. This may consist in perceiving only black and white for achromatopsia or slightly altered perception of red, green and blue for anomalous trichromacy (protanomaly, deuteranomaly, tritanomaly) or in a sever altered perception for dichromacy. In the following, we are focusing on dichromacy. The dichromacy is the result of the absence of one type of cone cells (among three) from eye retina responsible for trichromat vision. Many types of dichromacy exists: (1) deuteranopia (red-green deficiency due to the lack of M (green) cone cells), (2) protanopia (red-green deficiency - L (red) cone cells are missing) and (3) tritanopia (yellow-blue deficiency - S (blue) cone cells are absent).

Several algorithms to simulate dichromacy were developed [2,3,12]. The simulation algorithm proposed by Kuhn [12] is used for our experiments. The simulation for dichromacy proposed by the later is performed in the CIE L*a*b*, a color space which does not depend on the device on which the colors are represented. Many recoloring methods for dichromat were proposed for images [12,14,15], videos [13] and web pages [10,11].

2 Web Accessibility for CVD Users as an Optimization Problem

2.1 Contrast Loss on Textual Content

The color space used to represent colors on the Internet is sRGB (standard Red Green Blue). Let $u = (u^r, u^g, u^b)$ and v be two colors represented in the sRGB color space, where $u^i \in [0 : 255], i = \{r, g, b\}$. Let $L(u)$ be the luminance for the color u and $\Gamma_{u,v}$ the contrast between u and v according to [17]. We denote by $D(u) \in [0 : 255]^3$ the function that simulates dichromacy and $\Gamma_{u,v}^D$ the contrast ratio as perceived by a dichromat user. Then we have:

$$\Gamma_{u,v} = \frac{\max(L(u), L(v)) + 0.05}{\min(L(u), L(v)) + 0.05} \in [1 : 21] \tag{1}$$

and

$$L(u) = 0.2126 * h(u^r) + 0.7152 * h(u^g) + 0.0722 * h(u^b) \tag{2}$$

with

$$h(a) = \begin{cases} \frac{a/255}{12.92} & \text{if } a/255 \leq 0.03928 \\ \left(\frac{a/255 + 0.055}{1.055}\right)^{2.4} & \text{otherwise} \end{cases} \tag{3}$$

For the contrast ratio, several recommendations are made in WCAG 2.0. The guidelines 1.4.3 and 1.4.6 define minimum threshold at 4.5:1, respectively enhanced at 7:1 for the contrast ratio of textual information. For all contrast ratio values $\alpha \in [1 : 21]$ we have computed the maximal loss between the standard and the simulated contrast over the entire sRGB color space. The average maximal loss computed over the entire sRGB is important. It is at 1.9 for tritanope, 2.5 for protanope and 2.6 for deuteranope. The maximum contrast loss goes up to 3.8 for a protanope, 3.9 for a deuteranope and 3.7 for a tritanope.

2.2 Online Transformation of the Colors Through a User Hosted Proxy

We want to achieve on the fly transformation of the page using a user hosted proxy. This kind of proxy allows to perform specific transformation on any type of content even on secured content (HTTPs).

The experiments were performed with the help of Smart Web Accessibility Proxy (SWAP)[1]. SWAP is an open source project aiming to improve web accessibility and usability. The proxy part of the project allows, among others things to perform HTML and CSS analysis. The necessary time to carry out a specific transformation using the proxy is highly dependent on the user machine capabilities. So we need to improve the color contrast with varying time constraints. The goal is not the best solution but the best in the available time interval. It includes accessing the page (time variation), decoding the page and interpreting the colors (depends on the page size), improving colors (depend on the existing relationships between colors), recoding the page into HTML (depends on page size). CPU availability and user's computer capability are not controlled either.

Consequently, we have low control on the time required for a perfect page recoloring, so we must be able to interrupt the process at any time and achieve nevertheless good recoloring.

2.3 Modeling the Compensation as a Mono Objective Function

We denote by $\mathcal{C} = \{u_1, \ldots, u_{|\mathcal{C}|}\}$ the set of colors found on the page, and by $\mathcal{C}^F$ the corresponding set of transformed colors. We denote by $\mathcal{E} \subset \mathcal{C} \times \mathcal{C}$ the set of couples characterized by foreground and background colors found on the page. We denote by $\mathcal{C}^I$ the set of initial colors. Let be $\alpha \in [0, 1]$ a parameter that balances the importance of the contrast improvement versus color change and Δ_i the euclidean distance between the initial color u_i and the corresponding transformed color u_i^F in CIE L*a*b* color space. We define by $\Gamma_{i,j}^{F,D}$ the final contrast as perceived by a dichromat user for the couple of colors u_i and u_j. It is be given by: $\Gamma_{i,j}^{F,D} = \Gamma(D(u_i^F), D(u_j^F))$ where D is the simulation function and $\Gamma_{i,j}$ the contrast ratio as defined above. Our aim is to compensate the contrast loss and maintain in the same time a small amount of change in the colors. This conducts to the minimization of the following function:

$$F(u_1^F, u_2^F, \ldots, u_N^F) = (1-\alpha) \sum_{(u_i, u_j) \in \mathcal{E}} \frac{1}{2} \left[\max(\Gamma_{i,j}^I - \Gamma_{i,j}^{F,D}, 0) \right]^2 + \alpha \sum_{c_i \in \mathcal{C}} \frac{1}{2} \Delta_i^2 \quad (4)$$

where $\max(\Gamma_{i,j}^I - \Gamma_{i,j}^{F,D}, 0)$ guarantees that the contrast ratio of the final colors for a dichromat user is at least at the level of a standard user and Δ_i ensures that

---

[1] https://projectsforge.org/projects/swap/

the perceptual distance between the final and the initial colors is maintained and α is a constant used to weight between contrast compensation and reducing the change in colors.

3 CMA-ES for Color Compensation

3.1 CMA-ES

CMA-ES is a stochastic optimisation method for non-smooth and non-linear fitness function [9]. It is also one of the best evolutionary algorithm. In this work we evaluate the usability of CMA-ES for the contrast compensation problem for dichromat users. The method's working principle consists in searching a better solution (individual) among a set of candidate solutions (population of individuals) for a number of iterations (generations) [7]. We denote by $x_i^k \in \mathbb{R}^n, n \in \mathbb{N}$, the i-th individual at generation $k \in \mathbb{N}$, where $\eta \geq 2$ represents the population size and $\lambda \leq \eta$ the number of selected individuals from the population. At each generation, new individuals are sampled using multivariate normal distribution ($\mathcal{N}(\mu, \Sigma)$):

$$x_i^{k+1} \sim \sigma^k \mathcal{N}(\mu^k, \Sigma^k). \tag{5}$$

Also the mean, the covariation matrix and the step size are updated. The new mean is computed as a weighted average of the selected individuals sorted in a ascending order according to their fitness value. It is given by:

$$\mu^{k+1} = \sum_{i=1}^{\lambda} \omega_i x_i^{k+1}. \tag{6}$$

The covariation matrix is updated using an evolution path. An evolution path that exploits the information from the previous generation is built. Let $\lambda_{var} = \left(\sum_{i=1}^{\lambda} \omega_i^2 \right)^{-1}$ [7] be an indicator of the selection variance, $c_c \leq 1$ be the learning rate and $p_c^k \in \mathbb{R}^n$ be an evolution path at generation k. The new evolution path is given by:

$$p_c^{k+1} = (1 - c_c) p_c^k + \sqrt{c_c (2 - c_c) \lambda_{var}} \frac{\mu^{k+1} - \mu^k}{\sigma^k} \tag{7}$$

where $\frac{1}{c_c}$ is the back time horizon for p_c. At generation $k + 1$ the covariation matrix is given by:

$$\Sigma^{k+1} = (1 - c_1 - c_\lambda)\Sigma^k + c_1 p_c^{k+1} \left(p_c^{k+1} \right)^T + c_\lambda \sum_{i=1}^{\lambda} \omega_i z_i^{k+1} \left(z_i^{k+1} \right)^T \tag{8}$$

where $z_i^{k+1} = \frac{x_i^{k+1} - \mu^k}{\sigma^k}$, c_1 is approximately $\frac{2}{n^2}$ and c_λ is chosen to be around $\min \left(\lambda_{var}/n^2, 1 - c_1 \right)$. The step size update is given by [7]:

$$\sigma^{k+1} = \sigma^k exp\left(\frac{c_\sigma}{d_\sigma}\left(\frac{\|p_\sigma^{k+1}\|}{E\|\mathcal{N}(0,I)\|} - 1\right)\right) \tag{9}$$

where p_σ is a conjugate evolution path which at generation $k+1$ is built as follows:

$$p_\sigma^{k+1} = (1 - c_\sigma)\,p_\sigma^k + \sqrt{c_\sigma\,(2 - c_\sigma)\,\lambda_{var}}\,\left(\Sigma^k\right)^{-\frac{1}{2}}\frac{\mu^{k+1} - \mu^k}{\sigma^k}\,. \tag{10}$$

More details on the method can be found in $[1, 4\text{--}6, 8, 9]$.

3.2   Adaptation to Contrast Compensation

In our approach, we consider an individual i at generation $k = 0$ to be represented by the set of the original colors as follows: $x_i = (u_1^k, u_2^k, \ldots, u_N^k)$ where $u_i \in \mathcal{C}$ and $N = |\mathcal{C}|$. With the notation established in Sect. 2.3, we have designed the fitness function $F : [0 : 255]^{3N} \to \mathbb{R}^+$ given by (4). Minimizing F will provide a solution to our contrast compensation problem. By construction the fitness balances between contrast compensation and colors change. The contrast and the perceptual distance values are normalized in $[0 : 1]$ in the fitness expression to avoid scaling problems.

3.3   Real and Generated Dataset

In the analysis, we have considered two types of data. Real data $(\mathcal{D}_R)$ were obtained through CSS parsing of real web pages. For each page, a CSS analysis was performed in order to accurately extract the colors and the relationships between them. We consider a page as being fully represented by a set of entities characterized by foreground and background colors. The method is also tested on generated data $(\mathcal{D}_G)$. We have created a similar set of entities for which the contrast loss is artificially increased. To generate such colors, we have used the confusion line as defined in [3]. Overall data from over 350 pages real and generated ones were taken into account in our analysis. The real dataset has in average 9 colors and 8 distinct couples of colors per page. For each page, we have computed the number of couples of colors that display a contrast loss. The average percentages of those couples are presented in Fig. 1.

3.4   Experiments and Discussion

An experimental study was performed on both real and artificial data. The experimental study purpose is to identify the parameter settings that performs statisfactory after a small number of evaluations. As we are dealing with on-the-fly transformation, the ultimate goal is not as much as delivering a perfect solution but more like a satisfactory improvement in a short amount of time, the

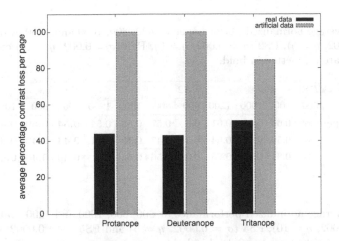

Fig. 1. Average percentage of color couples losing contrast

maximum time limit being defined by the user in the final application. The CMA-ES implementation, is the java version provided by N. Hansen[2]. This choice suites us, as tests were performed with the SWAP platform developped in java. Considering the experiments, 50 runs of the CMA-ES were performed for each page. Their average was considered as being the method result for that page.

Several hypothesis were tested. We have studied the relationship between the amount of compensation needed and the method behaviour (less compensation for real data, more compensation for artificial dataset). Also, we have tried to see if the method efficiency is bound to the initial step size and population size variation. A choice of parameter values was made. We have consider population size (η) of 5 and 10 individuals. We have considered the values $\sigma = \{0.02, 0.002, 0.0002\}$ and $\alpha = 0.15$. Considering α, higher values can be chosen but the choice for 0.15 was made under the hypothesis that a user may prefer a greater compensation of contrast than less variation in colors. We denote by $\mathcal{A}$ the set of all parameter settings obtained through the variation of the initial step size (σ) and population size (η) as mentioned above. For each parameter settings $a_i \in \mathcal{A}$ of CMA-ES, for each page p, the average fitness value for t evaluations is $F_p^{a_i}(t)$. We define:

$$f_p^m(t) = \min_{\substack{a_i \in \mathcal{A} \\ t=1..2500}} F_p^{a_i}(t), \ f_p^M(t) = \max_{\substack{a_i \in \mathcal{A} \\ t=1..2500}} F_p^{a_i}(t) \tag{11}$$

representing the minimum and the maximum fitness value over the parameter settings.

We were interested in establishing the global efficiency of the method. For this, we have computed the average normalized performance for each type of dataset (real and artificial) as follows:

---

[2] https://www.lri.fr/~hansen/CMA-ES_inmatlab.html#java

Table 1. Average normalized performance at 600, 800, 1000 and 1200 evaluations for PS1 ($\sigma = 0.02$, $\eta = 5$), PS2 ($\sigma = 0.002$, $\eta = 5$), PS3 ($\sigma = 0.002$, $\eta = 10$) on real data. Best values are typesetted in bold.

CVD type	PS1				PS2				PS3			
	600	800	1000	1200	600	800	1000	1200	600	800	1000	1200
Deuteranope	0.69	0.66	0.64	0.61	0.6	0.57	0.55	0.54	**0.54**	**0.50**	**0.48**	**0.47**
Protanope	0.63	0.59	0.56	0.54	0.5	0.47	0.46	0.44	**0.44**	**0.42**	**0.40**	**0.38**
Tritanope	0.84	0.81	0.78	0.75	0.59	0.56	0.54	0.51	**0.49**	**0.46**	**0.44**	**0.42**

Table 2. Average normalized performance at 600, 800, 1000 and 1200 evaluations for PS3 ($\sigma = 0.002$, $\eta = 10$), PS4 ($\sigma = 0.0002$, $\eta = 5$) and PS5 ($\sigma = 0.0002$, $\eta = 10$) on real data. Best values are typesetted in bold

CVD type	PS3				PS4				PS5			
	600	800	1000	1200	600	800	1000	1200	600	800	1000	1200
Deuteranope	**0.54**	**0.50**	**0.48**	**0.47**	0.67	0.63	0.6	0.57	0.67	0.61	0.57	0.53
Protanope	**0.44**	**0.42**	**0.40**	**0.38**	0.63	0.58	0.55	0.52	0.66	0.6	0.55	0.51
Tritanope	**0.49**	**0.46**	**0.44**	**0.42**	0.56	0.52	0.49	0.46	0.55	0.51	0.48	0.45

Table 3. Average normalized performance at 600, 800, 1000 and 1200 evaluations for PS1 ($\sigma = 0.02$, $\eta = 5$), PS2 ($\sigma = 0.002$, $\eta = 5$), PS3 ($\sigma = 0.002$, $\eta = 10$), PS4 ($\sigma = 0.0002$, $\eta = 5$), PS5 ($\sigma = 0.0002$, $\eta = 10$) on artificial data. Best values are typesetted in bold

CVD type	PS1				PS2				PS3			
	600	800	1000	1200	600	800	1000	1200	600	800	1000	1200
Deuteranope	0.01	0.01	0.01	0.01	0.02	0.01	0.01	0.01	**0.03**	**0.01**	**0.01**	**0**
Protanope	0.02	0.01	0.01	0.01	0.02	0.01	0.01	0.01	**0.03**	**0.01**	**0.01**	**0**
Tritanope	0.13	0.11	0.1	0.09	0.08	0.07	0.06	0.05	**0.05**	**0.04**	**0.04**	**0.03**

Table 4. Average normalized performance at 600, 800, 1000 and 1200 evaluations for PS3 ($\sigma = 0.002$, $\eta = 10$), PS4 ($\sigma = 0.0002$, $\eta = 5$) and PS5 ($\sigma = 0.0002$, $\eta = 10$) on artificial data. Best values are typesetted in bold.

CVD type	PS3				PS4				PS5			
	600	800	1000	1200	600	800	1000	1200	600	800	1000	1200
Deuteranope	**0.03**	**0.01**	**0.01**	**0**	0.42	0.29	0.2	0.15	0.57	0.4	0.29	0.21
Protanope	**0.03**	**0.01**	**0.01**	**0**	0.42	0.28	0.2	0.15	0.58	0.41	0.29	0.22
Tritanope	**0.05**	**0.04**	**0.04**	**0.03**	0.41	0.32	0.26	0.22	0.51	0.4	0.32	0.26

$$f_i^a(t) = \frac{1}{|\mathcal{D}_i|} \sum_{p \in \mathcal{D}_i} \frac{F_p^a(t) - f_p^m(t)}{f_p^M(t) - f_p^m(t)}, i \in \{R, G\}. \tag{12}$$

As we are dealing with an important time constraint, we are interested in determining the fitness behaviour after a small number of evaluations. The Tables 1, 2, 3 and 4 present the average normalized performance for all the parameter settings (PS) taken into account and for all types of datasets. This kind of knowledge is important as we are trying to be able to perform an on-the-fly transformation of the page. The chosen number of evaluation is not random, it may corresponds to user thresholds in the final application. A set of compensation levels may be proposed to the user from which he may choose the one that better suits his

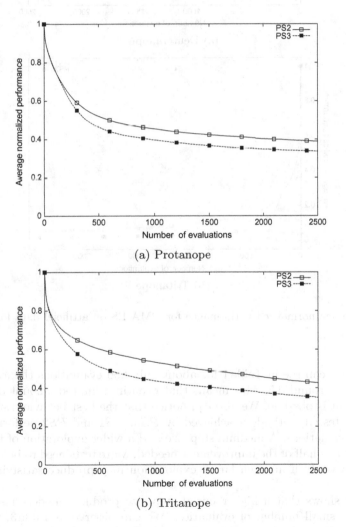

(a) Protanope

(b) Tritanope

Fig. 2. Average normalized performance for CMA-ES on real data for PS2 and PS3

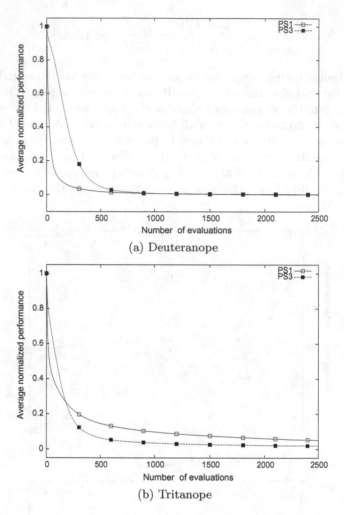

(a) Deuteranope

(b) Tritanope

Fig. 3. Average normalized performance for CMA-ES on artificial data for PS1 and PS3

needs. As we can see in Table 1 for about only 600 evaluations the average fitness reaches around 0.5. This means that even in a limited amount of time an improvement is possible. We also can notice that the best behaviour among the variants of tested methods is achieved by $PS3$. $PS1$ and $PS2$ perform modest after 600 evaluations. A medium step size and a wider exploration of the search space produces half of the improvement needed. More tests need to be performed in order to verify if an even larger exploration may produce satisfying results faster.

Table 2 shows that a higher initial step size produces modest results after a relatively small number of evaluation. We can observe in Table 3, that $PS3$ and $PS1$ perform better on pages in need of a higher level of compensation.

It can also be observed that protanope and deuteranope behaviour are similar for the choices of tested parameters. Also other parameters setting involving bigger initial step size may be tested.

We have measured the transformation time using the client proxy part of the SWAP project. The transformation includes: HTML parsing, CSS parsing, color extraction, building relationships between colors, colors optimization and sending the modified response with the optimized colors to the clients' browser.

A protanope transformation for a page with only 9 colors and 8 entities (which constitues the average in the real dataset used for experiments), needs about 4.3 s for 1200 evaluations using PS3 ($\sigma = 0.002$ and $\eta = 10$). For a page with 4 colors and 3 entities a tritanope transformation for 1200 evaluation takes about 1.3 s ($\sigma = 0.002$ and $\eta = 10$). Taking into account the statistics presented in Tables 1 and 3, we may conclude that our method can reach a satisfying compensation is a small amount of time. The mean for the best parameters setting was computed and presented in Figs. 2 and 3. Protanope and deuteranope have a similar behaviour for both real and artificial data. In general, CMA-ES tends to performs better when an important contrast compensation is needed, as seen in Fig. 3.

These results are highly dependent of our choices made considering the values of parameters and may be related to the considered datasets. A cluster analysis remains to be performed in order to establish the influence of the dataset characteristics as: number of colors on the page, the quantity of compensation needed, the complexity of the relationships between colors, the number of entities on the page that display a lower contrast. This set of parameters and much more need to be tested on the resulted clusters. This kind of analysis will provide useful knowledge which can be used in establishing appropriate thresholds in terms of number of evaluation for different types of parameters and pages. More work needs to be done on larger datasets.

4   Conclusion

In this paper, we propose an evolutionary approach to the contrast compensation for dichromat users. The performed experiments showed that CMA-ES works better when the contrast loss is important. The method behaviour is similar for protanope and deuteranope and it tends to perform slightly better for tritanope. Moreover, only about 700 evaluations are enough to obtain an acceptable solution to our problem. As time is an important aspect of our problem, reasonable amount of time maybe enough to obtain an improvement. Still more things remain to be done, for instance combination of parameters that help the method to converge faster when the compensation is little need to be found.

References

1. Auger, A., Hansen, N.: A restart CMA evolution strategy with increasing population size, vol. 2, pp. 1769–1776. IEEE. http://ieeexplore.ieee.org/lpdocs/epic03/wrapper.htm?arnumber=1554902
2. Brettel, H., Vienot, F., Mollon, J.: Computerized simulation of color appearance or dichromats. J. Opt. Soc. Am. **14**(10), 2647–2655 (1997)
3. Brettel, H., Vienot, F., Mollon, J.: Digital video colourmaps for checking the legibility of displays by dichromats. Color Res. Appl. **24**(4), 243–251 (1999)
4. Hansen, N., Auger, A., Ros, R., Finck, S., Posik, P.: Comparing results of 31 algorithms from the black-box optimization benchmarking BBOB-2009. In: 12th Annual Genetic and Evolutionary Computation Conference, pp. 1689–1696. ACM Press, New York (2010)
5. Hansen, N., Ostermeier, A.: Adapting arbitrary normal mutation distributions in evolution strategies: the covariance matrix adaptation, pp. 312–317. http://ieeexplore.ieee.org/lpdocs/epic03/wrapper.htm?arnumber=542381
6. Hansen, N.: Benchmarking a BI-population CMA-ES on the BBOB-2009 function testbed, p. 2389. ACM Press. http://portal.acm.org/citation.cfm?doid=1570256.1570333
7. Hansen, N.: The CMA evolution strategy: a comparing review **192**, 75–102. http://link.springer.com/10.1007/3-540-32494-1_4
8. Hansen, N., Ostermeier, A.: Completely derandomized self-adaptation in evolution strategies. Evol. Comput. **9**(2), 159–195 (2001). http://www.mitpressjournals.org/doi/abs/10.1162/106365601750190398
9. Hansen, N., Ros, R.: Black-box optimization benchmarking of NEWUOA compared to BIPOP-CMA-ES: on the BBOB noiseless testbed, p. 1519. ACM Press. http://portal.acm.org/citation.cfm?doid=1830761.1830768
10. Iaccarino, G., Malandrino, D., et al.: Efficient edge-services for colorblind users. In: WWW '06 The 15th International Conference on World Wide Web, Edinburgh, Scotland UK, pp. 919–920, 22–26 May 2006
11. Ichikawa, M., Tanaka, K., et al.: Web-page color modification for barrier-free color vision with genetic algorithm. In: GECCO'03 International Conference on Genetic and Evolutionary Computation: Part II, Chicago, USA, pp. 2134–2146 (2003)
12. Kuhn, G.R., Oliveira, M.M., Fernandes, L.A.F.: Efficient naturalness-preserving image-recoloring method for dichromats. IEEE Vis. Comput. Graph. **6**(14), 1747–1754 (2008)
13. Machado, G.M., Oliveira, M.M.: Real-time temporal-coherent color contrast enhancement for dichromats. Comput. Graph. Forum **29**(3), 933–942 (2010)
14. Park, J., Choi, J., Han, D.: Applying enhanced confusion line color transform using color segmentation for mobile applications, pp. 40–44. IEEE. http://ieeexplore.ieee.org/lpdocs/epic03/wrapper.htm?arnumber=5954274
15. Ruminski, J., Wtorek, J., Ruminska, J., Kaczmarek, M., Bujnowski, A., Kocejko, T., Polinski, A.: Color transformation methods for dichromats, pp. 634–641. IEEE. http://ieeexplore.ieee.org/lpdocs/epic03/wrapper.htm?arnumber=5514503
16. WCAG1. http://www.w3.org/TR/WCAG10/. Accessed November 2011
17. World Wide Web Consortium (W3C). http://www.w3.org/WAI/intro/components.php. Accessed September 2011

Multistart Evolutionary Local Search for a Disaster Relief Problem

Juan Carlos Rivera$^{(\boxtimes)}$, H. Murat Afsar, and Christian Prins

ICD-LOSI, Troyes University of Technology (UTT), 10000 Troyes, France
{rivera_j,murat.afsar,christian.prins}@utt.fr

Abstract. This paper studies the multitrip cumulative capacitated vehicle routing problem (mt-CCVRP), a variant of the classical capacitated vehicle routing problem (CVRP). In the mt-CCVRP the objective function becomes the minimization of the sum of arrival times at required nodes and each vehicle may perform more than one trip. Applications of this NP-Hard problem can be found in disaster logistics. This article presents a Multistart Evolutionary Local Search (MS-ELS) that alternates between giant tour and mt-CCVRP solutions, and uses an adapted split procedure and a variable neighborhood descent (VND). The results on two sets of instances show that this approach finds very good results in relatively short computing time compared with a multistart iterated local search which works directly on the mt-CCVRP solution space.

Keywords: Multi-trip cumulative capacitated vehicle routing problem · Disaster logistics · Evolutionary local search · Split · VND

1 Introduction

A recent trend is to apply operations research techniques to facilitate logistic operations in disaster relief. An important logistical issue after a disaster is to determine the transportation routes for first aids, supplies, rescue personnel or equipment between supply points and the destination nodes geographically scattered over the disaster region. The arrival time of relief supplies at the affected communities clearly impacts the survival rate of the citizens and the suffering.

In this sense, vehicle routing models can be considered for delivery in disaster context by using service-based objective functions to reflect the different priorities for delivering humanitarian aid (Campbell et al. [1]).

In this paper, the *multitrip cumulative capacitated vehicle routing problem* (mt-CCVRP) is studied. The mt-CCVRP (Rivera et al. [2]) is raised by the relief operations, in which (a) the classical objective function (total time or distance traveled) becomes the sum of arrival times at required nodes and (b) vehicles are allowed to perform multiple trips. This flexibility is necessary when the total demand exceeds the total capacity of the fleet of vehicles.

The paper is structured as follows: Sect. 2 briefly reviews the state of the art. In Sect. 3 the problem is formally defined. The proposed approach is developed in Sect. 4. Experimental results are presented in Sect. 5 and concluding remarks are given in Sect. 6.

© Springer International Publishing Switzerland 2014
P. Legrand et al. (Eds.): EA 2013, LNCS 8752, pp. 129–141, 2014.
DOI: 10.1007/978-3-319-11683-9_11

2   State of the Art

In relief context, it is critical that the deliveries to affected sites be both fast and fair. Campbell et al. [1] suggest that using service-based objective functions may better reflect the different priorities and strategic goals found in delivering humanitarian aid. The minimization of the average arrival times reflects better the emergency of humanitarian logistic operations than the classical objective functions such as the minimization of total tour length. Note that the minimization of the average arrival times is equivalent to the minimization of the sum of arrival times.

The sum of arrival times has been already used in some traveling salesman problems (TSP). For instance, the minimum latency problem consists in finding a tour starting at a depot and visiting each other node only once, in such a way that the total latency is minimized [3]. This problem is also known as the delivery man problem [4] or the travelling repairman problem (TRP) [5,6]. The multiple travelling repairman problem (k-TRP) generalizes the TRP where k tours must be determined [5].

The cumulative capacitated vehicle routing problem (CCVRP) is a variant of the CVRP where the objective function becomes the sum of arrival times at demand nodes. Ngueveu et al. [7] provide a mathematical model, several lower bounds and two memetic algorithms. Ribeiro and Laporte [8] present an adaptive large neighborhood search (ALNS) algorithm which is compared with the approach in [7], while Ke and Feng [9] improve some best known solutions with a two-phase metaheuristic by using exchange-based and cross-based operators to perturb the solutions in the first phase and local search moves in the second.

A comparison between cost minimization, minimization of the maximal arrival time and minimization of the average arrival times for the TSP and CVRP is given by Campbell et al. [1]. Their paper introduces lower bounds, an insertion heuristic and a local search procedure.

A common assumption is that each vehicle performs a single trip. Clearly, in many cases this assumption does not hold. Nevertheless, only Rivera et al. [2] consider multiple trips and the minimization of the sum of arrival times. These authors develop a non trivial mathematical model for mt-CCVRP, an MS-ILS metaheuristic and a dominance rule with respect to the order of trips in a multitrip.

Our approach uses a split procedure, proposed by Prins [10] for the CVRP and adapted by Ngueveu et al. [7] for the CCVRP. This procedure is adapted to mt-CCVRP, due to very particular features of the problem.

3   Problem Definition and Mixed Integer Linear Model

The problem is defined on an undirected complete graph $G = (V, E)$. The node-set $V = \{0, ..., n\}$ includes a depot-node 0 and a set $V' = V \setminus \{0\}$ of affected sites or required nodes. In the sequel, it is assumed that G is encoded as a symmetric directed graph. A fleet of R identical vehicles of capacity Q is based at the depot and each node $i \in V'$ has a known demand q_i. It is assumed without loss of generality that $\sum_{i \in V'} q_i \geq R \cdot Q$, $n \geq R$, and $q_i \leq Q$, $\forall i \in V'$.

The objective is to identify a set of trips such that each site is visited exactly once and the sum of arrival times at the sites is minimized. The important factor in the emergency relief operations is the arrival time at sites and not the total distance traveled by vehicles. A trip is defined as a circuit, starting and ending at the depot, in which the total demand serviced does not exceed vehicle capacity Q. Every trip must be assigned to exactly one vehicle and vehicles are allowed to perform more than one trip. The set of successive trips performed by one vehicle is called *multitrip*. The 0-1 mixed integer linear program (MILP), based on the model proposed by Rivera et al. [2], is defined by Eqs. (1)–(14).

$$\min\ Z = \sum_{i \in V} \sum_{j \in V} w_{ij} \cdot y_{ij} + \sum_{i \in V'} \sum_{j \in V'} (w_{i0} + w_{0j}) \cdot y'_{ij} \tag{1}$$

$$\sum_{i \in V'} x_{0i} = R \tag{2}$$

$$\sum_{i \in V'} (x_{ij} + x'_{ij}) + x_{0j} = 1, \ \forall\, j \in V' \tag{3}$$

$$\sum_{i \in V'} (x_{ji} + x'_{ji}) + x_{j0} = 1, \ \forall\, j \in V' \tag{4}$$

$$\sum_{j \in V} F_{ji} - \sum_{j \in V} F_{ij} = q_i, \ \forall\, i \in V' \tag{5}$$

$$F_{ij} \leq Q \cdot x_{ij}, \ \forall\, i,j \in V, i \neq j \tag{6}$$

$$F_{0j} \leq Q \cdot \left(x_{0j} + \sum_{i \in V'} x'_{ij} \right), \ \forall\, j \in V' \tag{7}$$

$$y_{ij} \leq (n - R + 1) \cdot x_{ij}, \ \forall\, i,j \in V, i \neq j \tag{8}$$

$$y'_{ij} \leq (n - R) \cdot x'_{ij}, \ \forall\, i,j \in V', i \neq j \tag{9}$$

$$\sum_{j \in V} (y_{ij} - y_{ji}) + \sum_{j \in V'} (y'_{ij} - y'_{ji}) = 1, \ \forall\, i \in V' \tag{10}$$

$$y_{ij} \geq 2 \cdot x_{ij} - x_{j0}, \ \forall\, i \in V, \ j \in V' \tag{11}$$

$$y'_{ij} \geq 2 \cdot x'_{ij} - x_{j0}, \ \forall\, i \in V', \ j \in V' \tag{12}$$

$$x_{ij} \in \{0, 1\}, \ y_{ij} \geq 0, \ F_{ij} \geq 0, \ \forall\, i,j \in V, i \neq j \tag{13}$$

$$x'_{ij} \in \{0, 1\}, \ y'_{ij} \geq 0, \ \forall\, i,j \in V', i \neq j \tag{14}$$

The model uses the concepts of replenishment arcs (Boland et al. [11]) and arc coefficients (Ngueveu et al. [7]). Variables are indexed by arcs and no trip nor multitrip index is required. F_{ij} define the flow (load) on each arc (i,j). The binary variables x_{ij} are equal to 1 if and only if arc (i,j) is traversed by a vehicle. Variables y_{ij} expresses the *arc coefficients* which are very useful to compute the objective function and prevent subtours. Arc coefficient y_{ij} designates the number of times the arc cost w_{ij} is counted in the solution cost. Similar variables

x'_{ij} and y'_{ij} are used for replenishment arcs. For instance, if a trip visits nodes 1, 2 and 3, the sum of arrival times is $(w_{01}) + (w_{01} + w_{12}) + (w_{01} + w_{12} + w_{23})$ and the arc coefficients are $y_{01} = 3$, $y_{12} = 2$ and $y_{23} = 1$.

The objective function (1) represents the sum of arrival times at affected sites. Constraints (2) mean that only R vehicles (R multitrips) can be used. Equations (3) and (4) respectively indicate that exactly one arc is traversed to arrive at site j and leave it. Constraints (5)–(7) concern flow variables and ensure that each demand is satisfied. Constraints (8)–(9) limit the number of arcs in a multitrip and Eq. (10) decrease the arc coefficients along trips.

Constraints (11) and (12) are new valid inequalities, which express that traversed arcs with not depot destination have an arc coefficient greater or equal than two, but becomes one if the destination of its immediate successor arc is the depot. Finally, constraints (13) and (14) define the five groups of variables.

4   Multistart Evolutionary Local Search

The proposed hybrid metaheuristic is a *multistart evolutionary local search* (MS-ELS). This method alternates between two kind of solution representations: mt-CCVRP solutions, which are sets of trips grouped in multitrips, and giant tours solutions without trip or multitrip delimiters. The metaheuristic also calls a *split* procedure and a *variable neighborhood descent* (VND) as improving phases. The proposed MS-ELS is sketched in Algorithm 1 while its internal components are described in the sequel.

A number (*MaxStart*) of successive Randomized Greedy Solutions are constructed. Every randomized initial solution S, is immediately improved by the VND. *Concatenate* procedure allows to translate the solution S in a giant tour T. After that, a number (*MaxIter*) of iterations are performed. In every iteration *MaxChildren* copies (T') of T are taken, a perturbation procedure (*Perturb*) is performed to each copy, the perturbed giant tours are optimally split up in multitrips (S'') by a *Split* procedure and improved by VND. The best of the *MaxChildren* solutions is used to replace S, in case of improvement. Finally, the best solution found S^* is updated when S improves the latter. The procedures *Precompute* and *Update* are used to speed up the VND.

In Algorithm 1, $Z(S^*)$ and $Z(S')$ define the global best cost and the cost of the best child of the current generation, respectively.

4.1   Solution Representations

In our algorithm, two solution representations are used. In a mt-CCVRP solution, noted by S, each multitrip is coded as a list of nodes in the order to be visited and uses the character "0" to delimit at the start and end of trips which means the visit to the depot. Giant tour solutions, noted by T, are composed of a single list of all required nodes in the order to be visited without trip or multitrip delimiters. The procedure *Concatenate* is used to transform a solution S to its correspondent T, while *Split* procedure optimally translates in $O(Rn^4)$ a giant tour T to an adequate S.

Algorithm 1. – MS-ELS

$Z(S^*) \leftarrow \infty$
for $Start \leftarrow 1$ **to** $MaxStart$ **do**
   $Greedy\_Randomized\_Heuristic(S)$
   $Precompute(\delta, S)$
   $VND(S)$
   for $Iter \leftarrow 1$ **to** $MaxIter$ **do**
      $Concatenate(S, T)$
      $Z(S') \leftarrow \infty$
      for $Child \leftarrow 1$ **to** $MaxChildren$ **do**
         $T' \leftarrow T$
         $Perturb(T')$
         $Split(T', S'')$
         $Update(\delta, S'')$
         $VND(S'')$
         if $Z(S'') < Z(S')$ **then**
            $S' \leftarrow S''$
         end if
      end for
      if $Z(S') < Z(S)$ **then**
         $S \leftarrow S'$
      end if
   end for
   if $Z(S) < Z(S^*)$ **then**
      $S^* \leftarrow S$
   end if
   if $Z(S) < Z(S^*)$ **then**
      $S^* \leftarrow S$
   end if
end for
$return\ S^*$

4.2   Pre-computations

Cost variation in moves is computed based on the concatenation operator $\oplus$, proposed by Silva et al. [12]. Given a sequence σ, $c(\sigma)$ denotes the cost to perform σ when starting at time 0, $t(\sigma)$ denotes the duration, $|\sigma|$ denotes the number of nodes in σ, and $\overleftarrow{\sigma}$ denotes the reversal of σ.

The cost and the duration of a sequence with one node are assumed to be 0 since there is no travel. The operator $\oplus$ concatenates two sequences, $\sigma = (u, ..., v)$ and $\sigma' = (u', ..., v')$. The following equations allow to compute the number of nodes, the duration and the cost values for $\sigma \oplus \sigma'$:

$$|\sigma \oplus \sigma'| = |\sigma| + |\sigma'| \tag{15}$$

$$t(\sigma \oplus \sigma') = t(\sigma) + w_{v,u'} + t(\sigma') \tag{16}$$

$$c(\sigma \oplus \sigma') = c(\sigma) + |\sigma'| \cdot (t(\sigma) + w_{v,u'}) + c(\sigma') \tag{17}$$

Note that the concatenation operator $\oplus$ is not commutative due to $c(\sigma \oplus \sigma') \neq c(\sigma' \oplus \sigma)$ and $t(\sigma \oplus \sigma') \neq t(\sigma' \oplus \sigma)$.

Moreover, moves browsed in VND often require reversals of sequences. Contrary to the routing problems with cost-based objective functions, the sum of arrival times of a sequence is modified after a reversal. In order to evaluate in $O(1)$ the cost variation of all moves, the cost for the reversal of any sequence σ in the incumbent solution must be prepared. The sequences considered contain required nodes only, so they must be completely contained in a trip.

Consider the reversal of $\sigma = \sigma' \oplus u$, where u is a single node and $\sigma' = (u', ..., v')$. The cost $c(\overleftarrow{\sigma})$ can be recursively deduced in $O(1)$ from the sequence σ' as $\overleftarrow{\sigma} = u \oplus \overleftarrow{\sigma'}$, using Eq. (18). Note that $c(\overleftarrow{\sigma'}) = 0$ if $|\sigma'| = 1$.

$$c(\overleftarrow{\sigma}) = c(v \oplus \overleftarrow{\sigma'}) = |\sigma'| \cdot w_{u,v'} + c(\overleftarrow{\sigma'}) \tag{18}$$

The reversal costs $c(\overleftarrow{\sigma})$ is computed by the *Precompute* procedure, before calling the VND after the initial solution, and stored in an array δ which is prepared in $O(n^2)$. *Update* procedure renovates the values of the reversal costs after the Split, but restricted to the sequences contained in modified multitrips.

4.3 Initial Solution Procedure

At each start of our MS-ELS, an initial solution is built by a greedy randomized heuristic. In this heuristic, each new trip is initialized with the farthest unserviced site. Then, a restricted candidate list (RCL) gathers the sites according with the following equation:

$$RCL = \{i \in V'' \mid z(i) \leq z_{min} + 0.05 \cdot (z_{max} - z_{min})\} \tag{19}$$

where V'' is the set of unserviced sites, $z(i)$ the insertion cost of site i in the emerging trip, and z_{max} and z_{min} the largest and smallest insertion costs. One site is randomly selected from the RCL to be added. The RCL is updated and sites are chosen until there is not enough capacity in the vehicle.

The trips are sorted in non-decreasing order of their mean duration, defined as total duration divided by its number of required nodes, then assigned to the end of the shortest partial multitrip. Rivera et al. [2] showed that the cost of a multitrip is minimized by ordering its trips in non-decreasing order of mean trip duration.

4.4 Variable Neighborhood Descent

The improvement procedure used in the MS-ELS is a variable neighborhood descent (VND), based on 8 neighborhoods. Each neighborhood is implicitly defined by a type of move. Starting from $p = 1$ and one input mt-CCVRP solution S, the basic iteration of VND consists in exploring the neighborhood N_p of S. As soon as a better solution is discovered, it replaces S and p is reset to 1, otherwise p is incremented. The procedure stops when the exploration of N_p brings no improvement. Only feasible solutions are accepted.

2-OPT moves on one trip (N_1): The 2-OPT move on one trip was already used by Ngueveu et al. [7] for the CCVRP. It consists in deleting two arcs and reconnecting the resulting fragments using two new arcs. Equivalently, it can be defined as the reversal of a sequence, represented as $\sigma_1 \oplus \overleftarrow{\sigma_2} \oplus \sigma_3$, where $\sigma_1 \oplus \sigma_2 \oplus \sigma_3$ is a trip. As a trip gets a different cost when inverted, we consider also the new variant represented as $\overleftarrow{\sigma_3} \oplus \sigma_2 \oplus \overleftarrow{\sigma_1}$.

λ-interchanges on one trip (N_2): Moves tested in this neighborhood consist in exchanging a sequence σ_1 from one to λ consecutive nodes with another (non-overlapping) sequence σ_2 containing zero to λ consecutive nodes. Each sequence with more than one node can be reversed in the reinsertion, giving four cases. Note that we allow a length of zero for the second sequence, to include relocations of the first string.

2-OPT moves involving two trips in a multitrip (N_3): This neighborhood takes two trips k and k' in the same multitrip, deletes one arc from each trip and reconnects them with different arcs. As we allow the reversal of each resulting sequence in these transformations, eight cases must be evaluated.

λ-interchanges involving two trips in a multitrip (N_4): Here, the λ-interchanges involve one sequence in each trip. As we allow the reversal of each sequence, four cases must be evaluated.

2-OPT moves on two trips done by distinct vehicles (N_5): 2-opt moves affecting two trips of different multitrips are similar to the ones browsed in N_3 but they use two trips in different multitrips. As both sequences come from different vehicles, a larger number of moves can be performed when $R \geq 2$.

λ-interchanges on two trips done by distinct vehicles (N_6): The λ-interchanges moves affecting two trips from two different multitrips are similar to the ones browsed in N_4. As in N_5, a larger number of moves can be performed when $R \geq 2$.

In all cases, cost variation of moves in N_1, N_2, N_3, N_4, N_5 and N_6 can be evaluated in constant time by using the concatenation operator $\oplus$. So, these neighborhoods can be browsed in $O(n^2)$.

Trip interchange (N_7): This neighborhood interchanges two trips from different multitrips. Every trip is inserted in the corresponding multitrip by following the dominance rule mentioned before. Note that only one order must be considered for every multitrip. As every move is evaluated in constant time, this neighborhood can be browsed in $O(R^2 n)$.

Trip splitting (N_8): This neighborhood adapts the route-splitting procedure for the multitrip VRP (Petch and Salhi [13]). This procedure starts by determining the shortest multitrip b, inspects each multitrip $m \neq b$ and evaluates the cost variation if the last trip of m is cut after the first, second, ..., last but one customer and moved at the end of multitrip b. This process can be implemented with an $O(Rn)$ complexity. In case of improvement, the VND goes back to neighborhood N_1, otherwise it terminates.

4.5 Concatenation

The concatenation procedure allows to transform a mt-CCVRP solution, S, to a giant tour solution, T, without trip or multitrip delimiters in a list of size n. The procedure starts by adding to an empty list, in order, the required nodes visited by the first multitrip, after that it continues adding the required nodes of the second multitrip, and so on, until multitrip R.

4.6 Perturbation

For each iteration, a perturbation procedure is applied to generate different children. This procedure consists in changing the direction of a sequence of nodes chosen randomly in a giant tour T. While this kind of perturbation changes the absolute position of a lot of nodes in the giant tour, the relative position remains similar. This kind of perturbation keeps together most of the nodes of the original solution but the new solution is different because (a) the cost of a sequence changes when it is reversed and (b) new trips are defined by the split procedure.

4.7 Split

This procedure allows to obtain a mt-CCVRP solution S from a giant tour solution T by solving a shortest path problem in an auxiliary digraph $H = (X, A, U)$. $X = \{0, 1, ..., n\}$ is the set of nodes. The arc set A contains all arcs $(i - 1, j)$ with $j > i$, where arc $(i - 1, j)$ means that sequence $\sigma = (T_i, ..., T_j)$ is visited by one vehicle. The mapping U defines the cost (sum of arrival times) $u_{i-1,j}$ of these arcs. Note that multitrips do not have capacity constraints.

The cost $u_{i-1,j}$ of arc $(i - 1, j) \in A$ can be computed by solving a supplementary shortest path problem on another auxiliary digraph $H' = (X', A', U')$, where the set of nodes X' contains a dummy node 0, and the nodes of sequence σ. The arc set A' contains arc $(i' - 1, j')$ if the sequence $\sigma' = (T_{i'}, ..., T_{j'})$ can be serviced by a feasible trip, respecting the capacity constraint. The set U' contains the cost $u'_{i'-1,j'}$ of arc $(i' - 1, j') \in A'$, which is the sum of arrival times of the corresponding trip.

Every time an arc $(i - 1, j) \in A$ is considered to compose a multitrip, two options are possible: either vehicles have enough capacity to visit all required nodes and the cost is computed as $u_{i-1,j} = u'_{i-1,j} = w_{0,T_i} + c(\sigma)$ or the multitrip must be split up in two or more trips. In the second case, the cost of arc $(i' - 1, j') \in A'$ can be computed as a function of the successive trips by using the Eq. (17). The paths on the graphs are built backward in order to deduce the splitting of the giant tour in $O(Rn^4)$.

5   Computational Experiments

5.1   Implementation

MS-ELS algorithm is implemented in Visual C++ and the mathematical model is solved by CPLEX 12.4. Both have been tested on a 2.50 GHz Intel Core i5 computer with 4 GB of RAM and Windows 7 Professional. Two kind of experiments are reported. The first one compares the solution of the 0-1 MILP via CPLEX solver with the hybrid metaheuristic. Such comparison is only possible on small instances. The second one compares the MS-ELS with the MS-ILS introduced by Rivera et al. [2].

Three sets of instances are used for our experiments: 12 small instances ($n = 15$) of Rivera et al. [2] (RAP), 7 modified instances of Christofides et al. [14] (CMT), and 20 modified instances of Golden et al. [15] (GWKC). Modifications consist in reducing the number of vehicles in order to force them to execute several trips and relaxing the trip length limit.

5.2   MS-ELS Parameter Tuning

The MS-ELS has only four parameters: the number of successive starts ($MaxStart$), the number of iterations per ELS ($MaxIter$), the number of children ($MaxChildren$) and the maximum number of consecutive sites in λ-interchange moves (λ). As the running time is roughly proportional to the number of calls to the VND, we decided to allocate a "computing budget" of 3000 calls to avoid excessive execution times. The best results on average are obtained with $MaxStart = 3$, $MaxIter = 100$, $MaxChildren = 10$ and $\lambda = 3$. For large instances results are also compared with $MaxStart = 3$, $MaxIter = 1000$, $MaxChildren = 1$ and $\lambda = 3$ which is equivalent to an MS-ILS by performing the split procedure.

5.3   Results on Small Instances

Table 1 compares the results for the 0-1 mixed integer linear program and the MS-ELS. The first four columns display the instance name, the number of required nodes, the number of vehicles and the average number of trips per vehicle $\sum_{i=1}^{n} q_i/(Q \cdot R)$. For MILP we provide linear relaxations, solution values (at the end of one hour, the best lower bound and the cost of the best integer solution found), the running times in seconds, and the percentage gap between best lower bounds and best cost of integer solutions. The best solution value over five runs, the average running time per run in seconds and the percentage gap between the best lower bound and the best cost are indicated for MS-ELS.

For the eight first instances, CPLEX finds an optimal solution. The instances look harder when the number of vehicles decreases and the average number of trips per vehicle increases: the running time of CPLEX augments quickly and the four last instances cannot be solved in one hour. However, the MS-ELS always returns a solution in at most 5.07 s (3 s on average). On the other hand, the

Table 1. Tests on small instances of Rivera et al. [2]

File	n	R	$\frac{\sum q_i}{Q \cdot R}$	0-1 MILP				MS-ELS (5 runs)		
				LR	Cost	Time	Gap	Best cost	Time	Gap
RAP_{01}	15	4	1.00	619.04	687.29	8.66	**0.00**	687.29	1.87	0.00
RAP_{02}	15	4	1.25	633.47	741.91	33.99	**0.00**	741.91	2.54	0.00
RAP_{03}	15	4	1.67	686.32	855.91	38.38	**0.00**	855.91	1.76	0.00
RAP_{04}	15	4	2.50	851.99	1090.67	24.20	**0.00**	1090.67	1.24	0.00
RAP_{05}	15	3	1.11	697.26	817.22	16.13	**0.00**	817.22	4.15	0.00
RAP_{06}	15	3	1.33	708.68	942.45	414.64	**0.00**	942.45	3.98	0.00
RAP_{07}	15	3	1.67	733.24	1008.03	560.17	**0.00**	1008.03	2.06	0.00
RAP_{08}	15	3	2.22	788.88	1111.04	124.54	**0.00**	1111.44	2.38	0.00
RAP_{09}	15	2	1.25	802.20	(1116.97/1182.66)	-	5.88	1182.66	5.07	5.88
RAP_{10}	15	2	1.67	814.44	(1100.45/1327.76)	-	20.65	1310.17	3.16	20.43
RAP_{11}	15	2	2.00	830.58	(1199.17/1391.60)	-	16.05	1391.60	3.28	16.05
RAP_{12}	15	2	2.50	862.40	(1296.97/1513.06)	-	16.66	1513.06	1.11	16.66
Mean						1301.72			3.07	

Table 2. Results for the instances of Christofides et al. [14]

Instance	n	R	$\frac{\sum q_i}{Q \cdot R}$	BKS	MS-ILS			MS-ELS[1]			MS-ELS[10]		
					D_{best}	D_{avg}	Time	D_{best}	D_{avg}	Time	D_{best}	D_{avg}	Time
CMT_{01}	50	3	1.62	3856.39	**0.00**	0.16	78.26	**0.00**	0.14	34.62	**0.00**	0.14	32.61
CMT_{02}	75	3	3.25	8300.15	**0.00**	0.04	68.37	**0.00**	0.01	47.59	**0.00**	0.03	48.22
CMT_{03}	100	3	2.43	10957.00	**0.00**	0.42	238.07	0.40	0.63	144.66	**0.00**	0.45	139.25
CMT_{04}	150	3	3.73	20595.93	0.01	0.25	479.34	**0.00**	0.07	316.62	**0.00**	0.21	329.71
CMT_{05}	199	3	5.31	33981.40	0.18	0.48	750.80	0.16	0.42	502.71	**0.00**	0.97	551.14
CMT_{11}	120	3	2.29	15797.40	**0.00**	0.17	470.41	**0.00**	0.11	217.54	**0.00**	0.14	275.24
CMT_{12}	100	3	3.02	10658.70	**0.00**	0.00	329.87	**0.00**	0.00	137.85	**0.00**	0.00	149.42
Average					0.03	0.22	345.02	0.08	0.20	200.23	**0.00**	0.28	224.46

MS-ELS reaches all optimal or best known solutions, and improves one of the best integer solutions found by CPLEX.

5.4   Results on Larger Instances

In this section, the proposed MS-ELS is compared with the MS-ILS of Rivera et al. [2] (without split). MS-ELS[10] refers to MS-ELS with ten children and 100 iterations while MS-ELS[1] refers the same method with one child and 1000 iterations. The results for larger instances are presented in Table 2 for CMT instances and Table 3 for GWKC instances, using the same columns: instance name, number of nodes n, number of vehicles R, average number of trips per vehicle $\sum_{i=1}^{n} q_i/(Q \cdot R)$, best known solution BKS, and for each method (MS-ILS, MS-ELS[1] and MS-ELS[10]) deviation from the best solution found in 5 runs in percent (D_{best}), average deviation of the 5 solutions from BKS in percent (D_{avg}) and average duration per run in seconds ($Time$).

On the CMT instances, the average computational time is less than 4 min, varying between 0.58 and 8.38 min for MS-ELS[1] and between 0.54 and 9.18 min

Table 3. Results for the instances of Golden et al. [15]

Instance	n	R	$\frac{\sum q_i}{Q \cdot R}$	BKS	MS-ILS			MS-ELS[1]			MS-ELS[10]		
					D_{best}	D_{avg}	Time	D_{best}	D_{avg}	Time	D_{best}	D_{avg}	Time
GWKC$_{01}$	240	4	2.18	122885.56	0.41	0.27	2443.14	0.52	0.29	1376.02	**0.00**	0.35	1530.17
GWKC$_{02}$	320	4	2.29	254443.84	0.06	0.23	5330.64	**0.00**	0.19	3624.22	**0.00**	0.22	3647.97
GWKC$_{03}$	400	4	2.22	434710.60	0.25	0.63	10123.85	0.77	0.83	8536.12	**0.00**	0.24	7779.81
GWKC$_{04}$	480	4	2.40	675615.36	0.16	0.44	15887.69	0.18	0.71	13478.36	**0.00**	0.66	14041.62
GWKC$_{05}$	200	3	1.48	193339.55	**0.00**	0.15	3018.47	0.07	0.43	1685.89	0.04	0.32	1799.27
GWKC$_{06}$	280	3	2.07	335034.88	**0.00**	0.22	5846.47	0.12	0.29	3381.19	0.02	0.26	3658.15
GWKC$_{07}$	360	3	2.67	493826.53	**0.00**	0.34	9144.21	0.15	0.47	7451.69	0.15	0.55	7301.47
GWKC$_{08}$	440	5	1.96	399995.84	**0.00**	0.42	13280.66	0.17	0.45	11153.84	0.51	0.60	11017.77
GWKC$_{09}$	255	5	2.69	13752.04	0.23	0.32	1433.82	**0.00**	0.39	993.54	0.35	0.51	1026.94
GWKC$_{10}$	323	6	2.53	18410.68	0.18	0.65	2374.63	**0.00**	0.52	1812.07	0.36	0.61	1905.67
GWKC$_{11}$	399	7	2.43	24239.20	0.13	0.33	3920.85	0.40	0.42	3565.39	**0.00**	0.57	3525.50
GWKC$_{12}$	483	8	2.34	31144.78	**0.00**	0.68	5831.21	0.08	0.61	5785.64	0.03	0.55	6210.10
GWKC$_{13}$	252	10	2.51	10056.24	0.25	0.27	913.49	0.13	0.36	492.32	**0.00**	0.27	541.42
GWKC$_{14}$	320	12	2.39	13413.98	0.32	0.25	1687.19	0.03	0.13	975.29	**0.00**	0.21	1031.94
GWKC$_{15}$	396	15	2.15	16385.79	0.14	0.16	2263.48	**0.00**	0.13	1884.62	0.14	0.19	1917.32
GWKC$_{16}$	480	15	2.39	24188.15	**0.00**	0.39	3737.54	0.25	0.86	3426.15	0.23	0.79	3320.24
GWKC$_{17}$	240	10	2.16	6618.18	**0.00**	0.63	791.04	0.29	0.68	433.94	0.25	0.62	443.48
GWKC$_{18}$	300	12	2.25	9385.02	**0.00**	0.43	1265.00	0.27	0.64	812.46	0.27	0.70	789.69
GWKC$_{19}$	360	12	2.70	14894.56	**0.00**	0.33	2032.59	0.72	0.89	1283.22	0.79	1.14	1352.93
GWKC$_{20}$	420	15	2.52	17939.90	**0.00**	0.49	2554.13	0.49	0.77	1946.51	0.64	1.05	2078.44
Average					0.11	0.38	4694.01	0.23	0.50	3654.92	0.19	0.52	3745.99

for MS-ELS[10]. MS-ELS[10] is 35 % faster than MS-ILS, while MS-ELS[1] is 42 % faster than MS-ILS.

With five runs, MS-ELS[10] always finds the best known solution, and two solutions are improved. The average costs for five runs is close to the best cost (0.28 %), indicating that MS-ELS is robust. MS-ELS[1] finds 5 best known solutions and improves two of the solutions finds by MS-ILS.

The best and average deviation for MS-ELS[10] increase moderately on the GWKC instances, with 0.19 % and 0.52 %, respectively. Nevertheless, 7 best known solutions are improved. The deviation from the best solution found (D_{best}) from MS-ELS[10] is larger than the D_{best} from MS-ILS due to the values of a few instances. But, if instances $GWKC_{08}$, $GWKC_{19}$ and $GWKC_{20}$ are ignored, D_{best} from MS-ELS[10] becomes 0.11 % while D_{best} from MS-ILS becomes 0.13 %.

MS-ELS[1] improves 4 best known solutions, one of them is also found by MS-ELS[10]. The best and average deviation for MS-ELS[1] are similar for both set of parameters.

The average computational time has been improved about 20 % by MS-ELS[1] and MS ELS[10], and it is ranging from 7 min to 234 min. The proposed meta-heuristic is still stable in terms of solution quality, yet the execution times vary a lot among instances of the same size. This variation is mainly due to the number of trips and the number of nodes per trip which have a great effect on the number of moves in neighborhoods.

A comparison between these methods shows that the use of the split procedure increases the efficiency of the MS-ELS algorithm respect to the MS-ILS, and the use of multiple children improves its performance.

6   Conclusions

The mt-CCVRP constitutes a good way to model the delivery of relief supplies after a humanitarian disaster, where the number of vehicles is limited and the time to reach affected areas is critical. The article presents an adapted split procedure and a VND in a hybrid multistart evolutionary local search algorithm.

On small instances, the resulting algorithm MS-ELS finds the same results as the mathematical model when the latter can be solved to optimality. The MS-ELS is able to produce competitive results in relatively short computing time. Some best known solutions has been improved for large instances.

A promising extension is the generalized CCVRP, in which the relief supplies must be delivered to one airport to be selected among the ones that are still operational in each region. Split deliveries should be allowed for a better use of vehicles.

References

1. Campbell, A.M., Vandenbussche, D., Hermann, W.: Routing for relief efforts. Transp. Sci. **42**(2), 127–145 (2008)
2. Rivera, J.C., Afsar, M.H.: Christian: A multi-start iterated local search for the multitrip cumulative capacitated vehicle routing problem. Technical report, Troyes University of Technology (2013)
3. Archer, A., Williamson, D.P.: Faster approximation algorithms for the minimum latency problem. In: Proceedings of the Fourteenth Annual ACM-SIAM Symposium on Discrete Algorithms (SODA) (2003)
4. Fischetti, M., Laporte, G., Martello, S.: The delivery man problem and cumulative matroids. Oper. Res. **41**(6), 1055–1064 (1993)
5. Jothi, R., Raghavachari, B.: Approximating the k-traveling repairman problem with repairtimes. J. Discrete Algorithms **5**(2), 293–303 (2007)
6. Tsitsiklis, J.N.: Special cases of traveling salesman and repairman problems with time windows. Networks **22**, 263–282 (1992)
7. Ngueveu, S.U., Prins, C., Calvo, R.W.: An effective memetic algorithm for the cumulative capacitated vehicle routing problem. Comput. Oper. Res. **37**(11), 1877–1885 (2010)
8. Ribeiro, G.M., Laporte, G.: An adaptive large neighborhood search heuristic for the cumulative capacitated vehicle routing problem. Comput. Oper. Res. **39**(3), 728–735 (2012)
9. Ke, L., Feng, Z.: A two-phase metaheuristic for the cumulative capacitated vehicle routing problem. Comput. Oper. Res. **40**, 633–638 (2013)
10. Prins, C.: A simple and effective evolutionary algorithm for the vehicle routing problem. Comput. Oper. Res. **31**(12), 1985–2002 (2004)
11. Boland, N., Clarke, L., Nemhauser, G.: The asymmetric traveling salesman problem with replenishment arcs. Eur. J. Oper. Res. **123**(2), 408–427 (2000)

12. Silva, M.M., Subramanian, A., Vidal, T., Ochi, L.S.: A simple and effective meta-heuristic for the minimum latency problem. Eur. J. Oper. Res. **221**(3), 513–520 (2012)
13. Petch, R.J., Salhi, S.: A multi-phase constructive heuristic for the vehicle routing problem with multiple trips. Discrete Appl. Math. **133**, 69–92 (2004)
14. Christofides, N., Mingozzi, A., Toth, P.: The Vehicle Routing Problem. Wiley, Chichester (1979)
15. Golden, B.L., Wasil, E., Kelly, J.P., Chao, I.M.: Metaheuristics in Vehicle Routing. Kluwer, Boston (1998)

Combinatorial and Discrete Optimization

Improving the Louvain Algorithm for Community Detection with Modularity Maximization

Olivier Gach[1,2]($\boxtimes$) and Jin-Kao Hao[2]

[1] LIUM, Université du Maine, Av. O. Messiaen, 72085 Le Mans, France
[2] LERIA, Université d'Angers, 2 Bd Lavoisier, 49045 Angers Cedex 01, France
olivier.gach@univ-lemans.fr, jin-kao.hao@univ-angers.fr

Abstract. This paper presents an enhancement of the well-known Louvain algorithm for community detection with modularity maximization which was introduced in [16]. The Louvain algorithm is a partial multi-level method which applies the vertex mover heuristic to a series of coarsened graphs. The Louvain+ algorithm proposed in this paper generalizes the Louvain algorithm by including a uncoarsening phase, leading to a full multi-level method. Experiments on a set of popular complex networks show the benefits induced by the proposed Louvain+ algorithm.

Keywords: Clustering · Optimization over networks · Heuristics

1 Introduction

Complex networks are a graph-based model which is very useful to represent connections and interactions of the underlying entities in a real networked system such as social [1], biological [2], and technological networks [3]. A vertex of the complex network represents an object of the real system while an edge symbolizes an interaction between two objects. For example in a social network, a vertex corresponds to a particular member of the network and an edge represents a relationship between two members. Complex networks typically display non-trivial structural and functional properties which impact the dynamics of processes applied to the network [4]. Analysis and synthesis of complex networks help discover these specific features, understand the dynamics of the networks and represent a real challenge for research [5,6].

A complex network may be characterized by a community structure. Vertices of a community are grouped to be highly interconnected while different communities are loosely associated with each other. Community is also called cluster or still module [7]. All the communities of a network form a clustering. In terms of graph theory, a clustering can be defined as a partition of the vertices of the underlying graph into disjoint subsets, each subset representing a community.

Intuitively, a community is a cohesive group of vertices that are more connected to each other than to the vertices in other communities. To quantify the

© Springer International Publishing Switzerland 2014
P. Legrand et al. (Eds.): EA 2013, LNCS 8752, pp. 145–156, 2014.
DOI: 10.1007/978-3-319-11683-9_12

quality of a given community and more generally a clustering, modularity is certainly the most popular measure [8]. Under this quality measure, the problem of community detection becomes a pure combinatorial optimization problem. Formally, the modularity measure can be stated as follows.

Given a weighted graph $G = (V, E, w)$ where w is a weighting function, i.e., $w : V \times V \longmapsto \mathbb{R}$ such that for all $\{u, v\} \in E, w(\{u, v\}) \neq 0$, and for all $\{u, v\} \notin E, w(\{u, v\}) = 0$. Let $X \subseteq V$ and $Y \subseteq V$ be two vertex subsets, $W(X, Y)$ the weight sum of the edges linking X and Y, i.e., $W(X, Y) = \sum_{u \in X, v \in Y} w(\{u, v\})$ (in this formula, each edge is counted twice). The modularity of a clustering with k communities $C = \{c_1, c_2, ..., c_k\}$ ($\forall i \in \{1, 2, ..., k\}, c_i \subset V$ and $c_i \neq \varnothing$; $\cup_{i=1}^{k} c_i = V; \forall i, j \in \{1, 2, ..., k\}, c_i \cap c_j = \varnothing$) is given by:

$$Q(C) = \sum_{i=1}^{k} \left[\frac{W(c_i, c_i)}{W(V, V)} - \left(\frac{d_i}{W(V, V)} \right)^2 \right] \quad (1)$$

where d_i is the sum of the degrees of the vertices of community c_i, i.e., $d_i = \sum_{v \in c_i} deg(v)$ with $deg(v)$ being the degree of vertex v.

It is easy to show that Q belongs to the interval [-0.5,1]. A clustering with a small Q value close to -0.5 implies the absence of real communities. A large Q value close to 1 indicates a good clustering containing highly cohesive communities. The trivial clustering with a single cluster has a Q value of 0.

Community detection with modularity is an important research topic and has a number of concrete applications [9]. In addition to its practical interest, community detection is also notable for its difficulty from a computational point of view. Indeed, the problem is known to be NP-hard [10] and constitutes thus a real challenge for optimization methods.

A number of heuristic algorithms have been proposed recently in the literature for community detection with the modularity measure. These algorithms follow three general solution approaches. First, greedy agglomeration algorithms like [11,12] iteratively merge two clusters that yield a clustering by following a greedy criterion. Second, local optimization algorithms like [13–15] improve progressively the solution quality by transitioning from a clustering to another clustering (often of better quality) by applying a move operator. The quality of such an algorithm depends strongly (among other things) on the move operator(s) employed. Third, hybrid algorithms like [16–19] combine several search strategies (e.g., greedy and multi-level methods) in order to take advantage of the underlying methods. Among the existing community detection algorithms, the Louvain algorithm presented in [16] (see next section) is among the most popular methods.

The Louvain algorithm belongs to the hybrid approach and can be compared to the general multi-level framework which requires both a coarsening and uncoarsening phases [20]. The coarsening phase reduces the size of a graph at each level by grouping several vertices of the original graph into a single vertex. The uncoarsening phase does the inverse by unfolding the vertices of the coarsen graph and then applying a refinement (optimization) procedure. While Louvain algorithm does use a coarsening phase, it omits the uncoarsening phase.

However, from an optimization point of view, it is known that the uncoarsening phase within the multi-level framework is useful to further improve the quality of the solution (see the example given in Sect. 2). This paper aims to extend the Louvain algorithm by including a uncoarsening phase, making the algorithm a full multi-level method. Experiments on a set of popular complex networks show the benefits induced by the proposed Louvain+ algorithm.

2 The Louvain Algorithm

The Louvain algorithm presented by Blondel et al. [16] operates on multiple levels of graphs, applying the vertex mover (VM) procedure on each level to improve the modularity. In this Section, we recall the two key elements of the methods: the VM procedure and the coarsening phase.

2.1 Vertex Mover Procedure

For a given graph where each vertex represents a community, one iteration of VM explores all the vertices of the graph in a random order. For each vertex, one examines all the possible moves to a neighbor community with an increased modularity. The move giving the largest increase is chosen and realized. At the end of an iteration, all the vertices of the current graph are processed. One proceeds with a new iteration if at least one vertex has migrated. To ensure that the vertices are examined in a purely random order during each iteration, the exploration of the vertices follows a random permutation of $\{1, 2, ..., n\}$ which is generated at the beginning once and for all. The procedure stops if no vertex has migrated when all the vertices have been examined. Another possible stop criterion is a minimum modularity gain: if the total gain obtained in one iteration is lower than the minimum gain required, the algorithm stops.

2.2 Coarsening Phase

The coarsening phase of the Louvain algorithm starts with the initial graph G (call it level 0 graph G^0) and produces a hierarchy of coarser graphs $G^1, G^2, ...$ of decreasing orders. We use $G^l = (V^l, E^l, w^l)$ to denote the graph of level l. From the graph G^0 and the initial trivial clustering where each vertex of G^0 forms a singleton community, the VM heuristic is applied to generate an improved clustering C^0. Then the graph G^1 of level 1 is created such that a vertex is introduced for each community of C^0 and an edge between two vertices is defined if they represent two neighboring communities in C^0. Now the VM heuristic is applied to the new graph G^1 with the clustering of singleton communities. This process continues and stops at some level L if the VM heuristic can not improve the initial clustering with singleton communities of G^L.

Formally, the generation of the coarsened graph G^{l+1} from (G^l, C^l) are achieved according to the following steps [21].

1. A vertex in G^{l+1} corresponds to a community of clustering C^l and vice versa. Given a community c of clustering C^l, let $T^{l+1}(c)$ denote the corresponding vertex in G^{l+1}.
2. Given two communities c and c' of clustering C^l, if they are connected by at least one edge in G^l, then their corresponding vertices $T^{l+1}(c)$ and $T^{l+1}(c')$ are linked by an edge in G^{l+1}. Additionally, the edge is weighted by $\frac{W^l(c,c')}{2}$.
3. A loop is added to each vertex $T^{l+1}(c)$ corresponding to community c weighted by $w^{l+1}(T^{l+1}(c), T^{l+1}(c)) = W^l(c,c)$.

This Louvain algorithm is illustrated on Fig. 1 with a simple graph containing 17 vertices and 29 edges.

3   Algorithm Louvain+

We extend the Louvain algorithm by introducing an uncoarsening-refinement phase at the end of the standard Louvain algorithm. Our Louvain+ algorithm executes the following steps:

1. Run the Louvain algorithm to obtain a series of coarsened graphs $G^1, G^2, ... G^L$ and clusterings $C^1, C^2, ... C^L$, assuming the highest level is L.
2. Run the uncoarsening phase from C^{L-1} and project the current clustering to a new clustering $\bar{C}^{L-2}$ where each coarsened community of the current clustering is unfolded (uncoarsened) into its composing communities. The new clustering $\bar{C}^{L-2}$ is immediately refined by the VM heuristic to improve its quality. The improved $\bar{C}^{L-2}$ serves then as the initial clustering for the next projection application. This process continues until level 0 is reached. (Notice that it is useless to start the uncoarsening phase from C^L since no moves are made by the VM heuristic during the last iteration of the coarsening phase.)

We describe now the process of projection. Given two vertices v_1^l and v_2^l of graph G^l, we use $v_1^l \, \Gamma^l \, v_2^l$ to denote the relation "v_1^l and v_2^l belong to the same community in C^l." Furthermore, we use $\gamma^l(v^l)$ to denote the community to which vertex v^l belongs in C^l. By convention, let $\bar{C}^{L-1} = C^{L-1}$ denote the first projected clustering. At each level $l = L - 2, L - 3...$, the clustering $\bar{C}^l$ is the result of the projection of $\bar{C}^{l+1}$ onto C^l which is optimized by the VM heuristic. In $\bar{C}^l$, two vertices v_1^l and v_2^l belong to the same community if the vertices in G^{l+1} corresponding to the communities $\gamma^l(v_1^l)$ and $\gamma^l(v_2^l)$ from C^l belong to the same community in C^{l+1}. Formally, this is denoted by $v_1^l \, \Gamma^l \, v_2^l \equiv T^l(\gamma^l(v_1^l)) \, \Gamma^{l+1} \, T^l(\gamma^l(v_2^l))$. This relation defines entirely the new clustering $\bar{C}^l$. The number of communities in $\bar{C}^{l+1}$ is the same as in $\bar{C}^l$. The uncoarsening phase with refinement by the VM heuristic is illustrated on Fig. 2 which starts with the result of Louvain algorithm (i.e., C^1) obtained in Fig. 1.

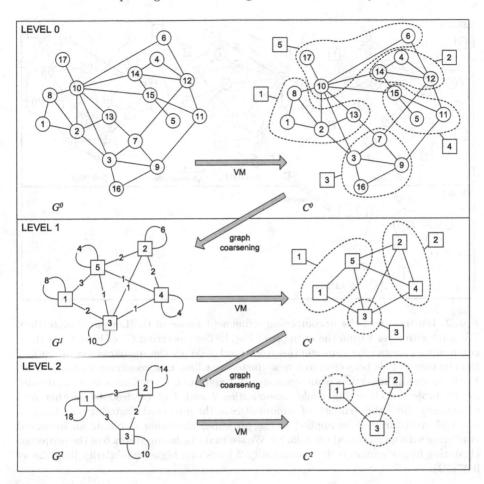

Fig. 1. Illustration of the Louvain algorithm. The initial graph G^0 contains 17 vertices and 29 edges. A first application of VM procedure to the trivial clustering of singleton communities gives the clustering C^0 composed of 5 communities. Then the coarsen graph G^1 is built with weighted edges and loops (squares in this graph represents communities from a lower level). The VM procedure is applied to the new graph G^1 to obtain the clustering C^1 with 3 communities. At level 2, the application of the VM procedure to the initial clustering of singleton communities does not change the clustering. The algorithm stops.

4 Experimental Results

4.1 Benchmark and Protocol of Test

To evaluate the efficiency of our Louvain+ algorithm, we compare it with the Louvain algorithm on a set of 13 networks from different application domains shown in Table 1. Both algorithms are coded in Free Pascal and executed on a

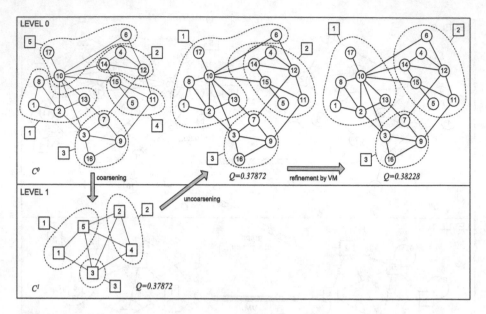

Fig. 2. Illustration of the uncoarsening-refinement phase of the Louvain+ algorithm. We start with level 1 from the example of Fig. 1. The clustering C^1 of level 1 has three communities containing communities from level 0. With the uncoarsening operation, the clustering C^1 is projected to a new clustering where new communities are formed. For instance all the vertices from communities 1 and 5 of C^0 now form a new community of the projected clustering while communities 2 and 4 of C^0 lead to another new community. Since the structure of communities in the projected clustering has changed, the VM procedure can be applied to the projected clustering to obtain an improved clustering with an increased modularity. We see that displacing vertex 6 of the projected clustering from community 1 to community 2 leads to a higher modularity (0.38228 vs 0.37872).

PC equipped with a Pentium Core i7 870 of 2.93 GHz and 8 GB of RAM[1]. Since the algorithm is sensitive to the order of vertices, we generate 100 instances of each graph with random vertices order. We use a deterministic version of the Louvain and Louvain+ algorithms (i.e. without preliminary random vertex reordering) and execute them on these 100 instances. For each graph, we present the distribution or average of different measures (modularity, number of vertices misplaced etc.) obtained over the 100 instances.

We use the minimal modularity gain ϵ between two consecutive iterations (see Sect. 2.1) as the stop condition of the VM procedure. We use ϵ_c and ϵ_r to distinguish the minimal modularity gain for the coarsening phase (for both Louvain and Louvain+) and for the uncoarsening phase (only Louvain+). It is

---

[1] The source code of our Louvain+ algorithm will be made available at www.info. univ-angers.fr/pub/hao/Louvainplus.html.

Table 1. Benchmark graphs in the literature for community detection with the number of vertices (n) and the number of edges (m). These are undirected graphs with medium size (from about 2000 to almost 1 million of edges).

Graph	Description	n	m	Source
Jazz	jazz musician collaborations network	198	2742	[22]
Email	university e-mail network	1133	5451	[23]
Power	topology of the Western States Power Grid of the United States	4941	6594	[24]
Yeast	Protein-Protein interaction network in yeast	2284	6646	[25]
Erdos	Erdös collaboration network	6927	11850	[26]
Arxiv	network of scientific papers and their citations	9377	24107	[27]
PGP	trust network of mutual signing of cryptography keys	10680	24316	[28]
Condmat2003	scientific coauthorship network in condensed-matter physics	27519	116181	[29]
Astro-ph	collaboration network of arXiv Astro Physics	16046	121251	[30]
Enron	email network from Enron	36692	183831	[31]
Brightkite	friendship network from a location-based social networking service	58228	214078	[32]
Slashdot	social network from Slashdot news web site	77359	469180	[31]
Gowalla	location-based social network from a website	196591	950327	[32]

clear that a smaller ϵ induces more applications of the VM heuristic and thus more computing time. In all of our experiments, we set $\epsilon_r = 10^{-5}$.

It is obvious that with the uncoarsening-refinement phase, the proposed Louvain+ algorithm will increase or leave unchanged the modularity which is achieved by Louvain. In the rest of this section, we assess experimentally the impact of the uncoarsening phase of Louvain+ on the run time cost, the modularity improvement and the structural changes of the clustering.

4.2 Execution Time and Modularity

Figure 3 shows a comparison of accumulated average runtime between Louvain and Louvain+ when they are applied to the set of 13 graphs with the same parameter value $\epsilon_c = \epsilon_r = 10^{-5}$. With the same coarsening phase in both algorithms, we can measure the extra time required by the uncoarsening-refinement phase of Louvain+. We observe that the curve of Louvain+ is slightly above that of Louvain but with a similar linear growth on m (number of edges in graph). The time complexity seems to be in $O(m)$. Curve *delta* shows a linear increase of runtime required by the uncoarsening-refinement phase. Louvain+ does not

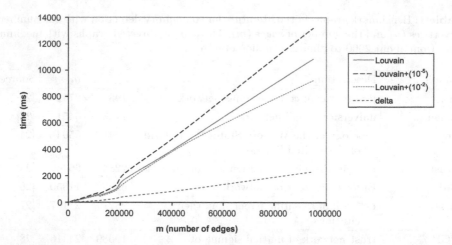

Fig. 3. Runtime comparison between Louvain and Louvain+. The three upper curves show the average run time of Louvain, Louvain+($\epsilon_c = 10^{-5}$) and Louvain+($\epsilon_c = 10^{-2}$) on the set of 13 graphs over 100 instances in milliseconds as a function of the number of edges m. The blue curve *delta* (lowest one) represents the time difference between Louvain+($\epsilon_c = 10^{-5}$) and Louvain (Color figure online).

change the complexity of the Louvain algorithm. Over the 13 tested graphs, the average increase of runtime caused by the refinement is about 20 %.

Figure 4 presents for each graph the gain of modularity given by the refinement of Louvain+. We observe that Louvain+ leads to an increase of modularity between 0.002 and 0.01 with respect to the results obtained by Louvain. This is achieved thanks to the uncoarsening phase introduced in Louvain+.

On the other hand, as shown in Fig. 3, Louvain+ consumes more CPU time than Louvain to achieve the reported (better) results. One interesting question is to know whether Louvain+ is able to attain the same results with less computing time. To verify this, we carry out another experiment where we run Louvain+ with a relaxed coarsening phase by using a much larger ϵ_c value ($\epsilon_c = 10^{-2}$ instead of $\epsilon_c = 10^{-5}$).

Now observe again Fig. 4 for the modularity gain of Louvain+. It can be seen that Louvain+ with $\epsilon_c = 10^{-2}$ leads to a modularity performance comparable to that with $\epsilon_c = 10^{-5}$ while the computing time is decreased, and becomes lower than the computing time of Louvain. This can be explained as follows. With the relaxed ϵ_c value, the coarsening phase is reduced. Even if this generally leads to a clustering with a decreased modularity at the end of the coarsening phase, the modularity is improved during the uncoarsening-refinement phase.

4.3   Bad Vertices and Structural Changes in Clustering

We now turn our attention to evaluate the structural changes in clustering made by the uncoarsening-refinement phase of Louvain+. For this purpose, we compare the clusterings obtained before and after the uncoarsening-refinement phase,

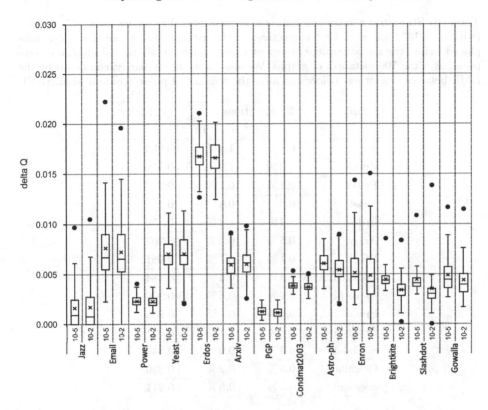

Fig. 4. Box-and-Whisker Plots for the modularity gain obtained by the uncoarsening-refinement phase on 100 instances of each graph. Two versions are tested, one with $\epsilon_c = 10^{-5}$ and one with $\epsilon_c = 10^{-2}$ (parameter value on x-axis). In both cases, the final expected precision is the same, $\epsilon_r = 10^{-5}$.

corresponding to the results of Louvain and Louvain+ respectively. An interesting measure for this evaluation is the percentage of misplaced vertices according to the strong sense of community criterion [33] correctly placed by the refinement. A community is defined in the strong sense if the internal degree of all the vertices of the community is greater than the external degree (there are more adjacent vertices in a community than outside). This is a very strong condition of existence of a community which is rarely satisfied in real networks, but it is interesting to count the number of vertices that do not satisfy this condition for a given clustering. To simplify our discussion, we use the term 'correction' to designate these vertices misplaced by Louvain (i.e., those vertices with an internal degree smaller than some external degree), but correctly placed by Louvain+, i.e. by the refinement phase. Generally, according to our observations, the maximum of modularity goes with the minimum of misplaced vertices.

We show in Table 2 the percentage of vertices corrected by the refinement phase of Louvain+ and the similarity between clusterings before and after this

Table 2. Structural comparison between Louvain and the two versions of Louvain+ (with $\epsilon_c = 10^{-5}$ and $\epsilon_c = 10^{-2}$ respectively). The average percentage of vertices, misplaced before the uncoarsening-refinement phase and correctly placed after, are computed over the 100 instances of graphs. We also show the similarity, computed by the NMI, between clusterings before and after the refinement phase (column 'similarity').

Graph	% corrections		Similarity
	$\epsilon_c = 10^{-5}$	$\epsilon_c = 10^{-2}$	
Jazz	0.5 %	1.3 %	0.953
Email	1.2 %	1.2 %	0.879
Power	0.1 %	0.1 %	0.947
Yeast	0.6 %	0.4 %	0.869
Erdos	1.4 %	1.4 %	0.856
Arxiv	0.5 %	0.4 %	0.907
PGP	0.1 %	0.1 %	0.981
Condmat2003	0.4 %	0.4 %	0.884
Astro-ph	0.7 %	0.5 %	0.906
Enron	0.2 %	0.1 %	0.955
Brightkite	0.3 %	0.1 %	0.931
Slashdot	0.3 %	0.0 %	0.845
Gowalla	0.3 %	0.2 %	0.930
average	0.5 %	0.5 %	0.911

phase. We find that the percentage of corrections is positive for all the tested graphs. This percentage represents 0.1 % to 1.4 % of the total vertices, with an average of 0.5 % over all the tested graphs. This information allows us to confirm once again the usefulness of the uncoarsening-refinement phase introduced in the Louvain+ algorithm.

We also calculate the global structural difference between clusterings before and after the refinement phase, measured by the similarity called NMI [34]. This measure is based on information theory and mostly used in community detection. The range of NMI goes from 0 (completely different clusterings) to 1 (identical clusterings). Table 2 discloses that structural changes made by the Louvain+ refinement is quite important with a NMI between 0.84 and 0.98. As the NMI scale is logarithmic, a value of 0.9 implies a significant structural difference.

5   Conclusion and Perspectives

In this work, we have presented an improved algorithm for community detection with modularity. The proposed Louvain+ algorithm extends the well-known Louvain algorithm by adding an uncoarsening-refinement phase, leading to a fully multi-level method. From the result of the Louvain algorithm, this extension goes backward and uncoarsens successively each intermediate graph generated during the Louvain algorithm and applies the vertex mover heuristic to each uncoarsened graph to improve the modularity. We have assessed the performance of the proposed algorithm on a set of 13 popular real networks. The comparisons with Louvain show that with comparable computing times, Louvain+ achieves systematically better modularity than Louvain does, thanks to the optimization during the uncoarsening-refinement phase. Experiments also disclosed that the extension of the uncoarsening phase does not change the linear complexity of the initial Louvain algorithm.

Like Louvain, the proposed Louvain+ algorithm is conceptually simple and computationally fast. As a consequence, it can be applied to very large networks that can be encountered in numerous real situations. Additionally, it can be used within more sophisticated methods, e.g. to generate initial clusterings that are further improved by search-based heuristics.

Acknowledgment. We are grateful to the referees for their comments and questions which helped us to improve the paper. The work is partially supported by the Pays de la Loire Region (France) within the RaDaPop (2009–2013) and LigeRO (2010–2013) projects.

References

1. Girvan, M., Newman, M.E.J.: Community structure in social and biological networks. Proc. Natl. Acad. Sci. USA **99**(12), 7821–7826 (2002)
2. Guimerà, R., Amaral, L.A.N.: Functional cartography of complex metabolic networks. Nat. **433**(7028), 895–900 (2005)
3. Flake, G.W., Lawrence, S., Giles, C.L., Coetzee, F.M.: Self-organization and identification of web communities. Comput. **35**(3), 66–70 (2002)
4. Boccaletti, S., Latora, V., Moreno, Y., Chavez, M., Hwang, D.: Complex networks: structure and dynamics. Phys. Rep. **424**(4–5), 175–308 (2006)
5. Strogatz, S.H.: Exploring complex networks. Nat. **410**(6825), 268–276 (2001)
6. Albert, R., Barabási, A.-L.: Statistical mechanics of complex networks. Rev. Mod. Phys. **74**, 47 (2002)
7. Newman, M.: Networks: An Introduction. Oxford University Press, New York (2010)
8. Newman, M.E.J., Girvan, M.: Finding and evaluating community structure in networks. Phys. Rev. E **69**(2), 026113 (2004)
9. Fortunato, S.: Community detection in graphs. Phys. Rep. **486**(3–5), 75–174 (2010)
10. Brandes, U., Delling, D., Gaertler, M., Gorke, R., Hoefer, M., Nikoloski, Z., Wagner, D.: On modularity clustering. IEEE Trans. Knowl. Data Eng. **20**(2), 172–188 (2008)
11. Clauset, A., Newman, M.E.J., Moore, C.: Finding community structure in very large networks. Phys. Rev. E **70**(6), 066111 (2004)
12. Newman, M.E.J.: Fast algorithm for detecting community structure in networks. Phys. Rev. E **69**(6), 066133 (2004)
13. Schuetz, P., Caflisch, A.: Efficient modularity optimization by multistep greedy algorithm and vertex mover refinement. Phys. Rev. E **77**(4), 046112 (2008)
14. Lü, Z., Huang, W.: Iterated tabu search for identifying community structure in complex networks. Phys. Rev. E **80**(2), 026130 (2009)

15. Duch, J., Arenas, A.: Community detection in complex networks using extremal optimization. Phys. Rev. E **72**(2), 027104 (2005)
16. Blondel, V.D., Guillaume, J.-L., Lambiotte, R., Lefebvre, E.: Fast unfolding of communities in large networks. J. Stat. Mech. Theor. Exp. **10**, P10008 (2008)
17. Liu, X., Murata, T.: Advanced modularity-specialized label propagation algorithm for detecting communities in networks. Phys. A **389**(7), 1493–1500 (2009)
18. Noack, A., Rotta, R.: Multi-level algorithms for modularity clustering. In: Vahrenhold, J. (ed.) SEA 2009. LNCS, vol. 5526, pp. 257–268. Springer, Heidelberg (2009)
19. Gach, O., Hao, J.-K.: A memetic algorithm for community detection in complex networks. In: Coello, C.A.C., Cutello, V., Deb, K., Forrest, S., Nicosia, G., Pavone, M. (eds.) PPSN 2012, Part II. LNCS, vol. 7492, pp. 327–336. Springer, Heidelberg (2012)
20. Walshaw, C.: Multilevel refinement for combinatorial optimisation problems. Ann. Oper. Res. **131**, 325–372 (2004)
21. Arenas, A., Duch, J., Fernández, A., Gómez, S.: Size reduction of complex networks preserving modularity. N. J. Phys. **9**(6), 176 (2007)
22. Gleiser, P., Danon, L.: Community structure in social and biological networks. Adv. Complex Syst. **6**, 565–573 (2003)
23. Guimerà, R., Danon, L., Díaz-Guilera, A., Giralt, F., Arenas, A.: Self-similar community structure in a network of human interactions. Phys. Rev. E **68**(6), 065103 (2003)
24. Watts, D.J., Strogatz, S.H.: Collective dynamics of "small-world" networks. Nat. **393**(6684), 440–442 (1998)
25. Bu, D., Zhao, Y., Cai, L., Xue, H., Zhu, X., Lu, H., Zhang, J., Sun, S., Ling, L., Zhang, N., Li, G., Chen, R.: Topological structure analysis of the protein-protein interaction network in budding yeast. Nucleic Acids Res. **31**(9), 2443–2450 (2003)
26. Grossman, J.: The erdös number project (2007). http://www.oakland.edu/enp/
27. KDD, Cornell kdd cup (2003). http://www.cs.cornell.edu/projects/kddcup/
28. Boguñá, M., Pastor-Satorras, R., Díaz-Guilera, A., Arenas, A.: Models of social networks based on social distance attachment. Phys. Rev. E **70**(5), 056122 (2004)
29. Newman, M.E.J.: The structure of scientific collaboration networks. Proc. Natl. Acad. Sci. USA **98**(2), 404–409 (2001)
30. Leskovec, J., Kleinberg, J., Faloutsos, C.: Graph evolution: densification and shrinking diameters. ACM Trans. Knowl. Disc. Data **1**(1), Article No. 2 (2007). doi:10.1145/1217299.1217301
31. Leskovec, J., Lang, K.J., Dasgupta, A., Mahoney, M.W.: Community structure in large networks: natural cluster sizes and the absence of large well-defined clusters. Internet Math. **6**(1), 66 (2008)
32. Cho, E., Myers, S.A., Leskovec, J.: Friendship and mobility: user movement in location-based social networks. In: Proceedings of the 17th ACM SIGKDD International Conference on Knowledge Discovery and Data Mining, pp. 1082–1090 (2011)
33. Radicchi, F., Castellano, C., Cecconi, F., Loreto, V., Parisi, D.: Defining and identifying communities in networks. Proc. Natl. Acad. Sci. USA **101**(9), 2658–2663 (2004)
34. Danon, L., Díaz-Guilera, A., Duch, J., Arenas, A.: Comparing community structure identification. J. Stat. Mech. Theor. Exp. **2005**(09), P09008 (2005)

A Recombination-Based Tabu Search Algorithm for the Winner Determination Problem

Ines Sghir[1,2], Jin-Kao Hao[1](✉), Ines Ben Jaafar[2], and Khaled Ghédira[2]

[1] LERIA, Université d'Angers, 2 Bd Lavoisier, 49045 Angers Cedex 01, France
hao@info.univ-angers.fr
[2] SOIE, ISG, Université de Tunis, Cité Bouchoucha, 2000 Le Bardo, Tunis, Tunisia
{inessghir,ines.benjaafar}@gmail.com,
khaled.ghedira@isg.rnu.tn

Abstract. We propose a dedicated tabu search algorithm (TSX_WDP) for the winner determination problem (WDP) in combinatorial auctions. TSX_WDP integrates two complementary neighborhoods designed respectively for intensification and diversification. To escape deep local optima, TSX_WDP employs a backbone-based recombination operator to generate new starting points for tabu search and to displace the search into unexplored promising regions. The recombination operator operates on elite solutions previously found which are recorded in an global archive. The performance of our algorithm is assessed on a set of 500 well-known WDP benchmark instances. Comparisons with five state of the art algorithms demonstrate the effectiveness of our approach.

Keywords: Winner determination problem · Tabu search · Solution recombination · Combinatorial optimization · Heuristics

1 Introduction

An auction involves an auctioneer wishing to maximize his/her selling revenue and a set of bidders wishing to minimize their cost according to their valuations of the items that they want to acquire. Examples of the most widely known auctions are the English auction, the Holland's auction, the Sealed envelope auction, and the Vickrey auction [12]. These auctions typically handle one item per sell.

Combinatorial auctions are multi-item auctions, which allow bids on combinations of items [5,11]. In a combinatorial auction, we are given a set of items exposed to buyers. Buyers offers different bids, each bid being defined by a subset of items with a price (bidder's valuation). Two bids are conflicting if they share at least one item. The Winners Determination Problem (WDP) is to determine a conflict-free allocation of items to bidders (the auctioneer can keep some of the items) that maximizes the auctioneer's revenue defined as the sum of the valuations of the winning bids [14]. The WDP is known to be a NP-hard problem with a number of practical applications like e-commerce, games theory and resources allocation in multi-agents systems [11,21].

© Springer International Publishing Switzerland 2014
P. Legrand et al. (Eds.): EA 2013, LNCS 8752, pp. 157–167, 2014.
DOI: 10.1007/978-3-319-11683-9_13

Formally, given a set of items $M = \{1, 2, ..., m\}$ and a set of n bids $N = \{1, 2, ...n\}$. Each bid j is a tuple $<S_j, P_j>$ where S_j is a subset of items covered by bid j, and P_j, the price of bid j. Let B be a $m \times n$ binary matrix such that $B_{ij} = 1$ if object $i \in S_j$, $B_{ij} = 0$ otherwise. Furthermore, define a decision variable x_j for each bid j such that $x_j = 1$ if bid j is a winning bid, 0 otherwise. Then, the WDP can be stated as the following binary integer optimization problem.

$$Maximize \; f(x) = \sum_{j \in N} P_j x_j \qquad (1)$$

subject to

$$\sum_{j \in N} B_{ij} x_j \leq 1, i \in M \qquad (2)$$

The objective function (1) allows to maximize auctioneer's gain calculated by the sum of prices of the winning bids while the constraints expressed by formula (2) ensure that an item appears at most in one winning bid.

The computational challenge of the WDP and its practical applications have motivated a number of solution approaches including exact methods [18] and metaheuristic methods. Representative examples of exact methods include: Branch-on-Items (BoI), Branch-on-Bids (BoB) [19], Combinatorial Auctions BoB (CABoB) [20], Combinatorial Auction Structural Search (CASS) [6] and Combinatorial Auctions Multi-unit Search (CAMUS) [15]. A dynamic programming approach is introduced in [17] while a linear programming method is investigated in [16]. An algorithm based on integer programming is shown in [1], a constraint programming approach is used to solve a particular combinatorial Vickrey auction [9]. On the other hand, several stochastic methods were proposed for the WDP. They include a local search method named Casanova [10], a hybrid algorithm combining simulated annealing with Branch-and-Bound (SAGII) [8], and more recently a tabu search method [3] and a memetic algorithm [4].

The rest of the paper is organized as follows. Section 2 describes the proposed algorithm which is based on two complementary neighborhoods and a recombination operator. Experimental results are reported in Sect. 3 and compared with five representative algorithms for the WDP. Finally, Sect. 4 concludes the paper.

2   Recombination-Based Tabu Search for the WDP

TSX_WDP uses two complimentary move operators to explore effectively the search space and a recombination operator as an additional means to escape deep local optima. In this section, we presents in detail these key components.

2.1   The Solution Representation

A candidate solution is represented by an allocation A (a dynamic vector). Each element of this allocation A receives the winning bid. Each bid is an object composed of the list of items and the associated prices.

2.2   The Evaluation Function

The objective function defined in Eq. (1) is used to measure the quality of a candidate solution. So if an allocation A contains k bids $\{B_1, B_2, ..., B_k\}$, ($B_i = <S_i, P_i>, 1 \leq i \leq k, k \leq n$), its quality is just equal to $f(A) = \sum_{i=1}^{k} P_i$, i.e., the sum of the valuations of the winning bids. Given two candidate solutions, the one with a higher objective value is considered to be better. This relation is used to compare neighboring solutions which are developed below.

2.3   The Basic Move Operators and the Neighborhoods

Our TSX_WDP algorithm explores the search space by using two complementary neighborhood relations which are defined by an intensification move operator and a perturbation move operator.

Intensification Move. The intensification move operator chooses bids among candidate bids to be inserted in the current allocation A. During one iteration of the algorithm, several bids can be selected if they improve the current allocation. To create a neighboring allocation, the following steps are followed:

- The initial candidate bids are sorted according to their utility prices;
- For each candidate bid B_x, a binary gain function is used to verify if the bid can increase the revenue of the current allocation when the bid is inserted;
- Let Q be the set of winning bids that are in conflict with the current candidate bid B_x, Let $f(Q)$ be the revenue of the set of winning bids Q, and $f(B_x)$ the price of the candidate bid B_x. The gain function returns true if $f(Q) < f(B_x)$ and returns false otherwise;
- Based on the function f, a candidate bid B_x can be added to the current allocation only if its price $f(B_x)$ is higher than the revenue of other winning bids which are conflicting with B_x in the current allocation (i.e., the gain function is true);
- The gain of B_x, when it is selected to be added in the current allocation, is calculated by: $Gain(B_x) = f(A) - f(Q) + f(B_x)$;
- When a bid B_x is inserted in the current allocation A, the bids of Q which are conflicting with B_x are removed from A;
- The steps mentioned previously are iterated until all the initial candidate bids are visited and possibly added in the current allocation A.

Perturbation Move. The perturbation move operator chooses randomly one candidate bid from the available ones. This move is activated only if no bid among the candidate bids can improve the current solution. In fact, the application of the intensification move can make the search to be trapped into local optima during the search process, when no more bid can be found that improves the revenue of the current allocation. Notice that this move operator can decrease temporarily the revenue of the solution, but hopefully, it helps the search to escape local optima by displacing the search to new zones of the search space. This move operator plays thus a diversification role.

2.4   Tabu List and Tabu Tenure Management

Tabu search uses a tabu list to forbid recently visited solutions from being revisited. The TSX_WDP algorithm considers the following general prohibition rule: a bid that is chosen to be inserted in the current allocation A (by an intensification move or a perturbation move) is forbidden to be removed for the next tt iterations (called tabu tenure). tt is calculated dynamically by the function proposed in [7]: $tt = L + \lambda + f(A)$ where L is randomly chosen from the interval $[0, 9]$ and λ is empirically fixed to 0.6. Experimentations show this dynamic tabu tenure is robust and allows TSX_WDP to reach high quality solutions. Notice that we permit a move to be accepted in spite of being tabu if the move leads to a solution better than any found so far. This is called the aspiration criterion.

2.5   Elite Solution Archive

The proposed algorithm also relies on a solution recombination operator (see next section) which aims to blend elite solutions (high-quality local optima). This technique is based on an archive P which is built as follows. During the search, if the current best solution A^* is not improved within a fixed number p of consecutive iterations, A^* is considered as a good local optimum and is added into the archive P. At the same time, this allocation corresponds to a deep local optimum which is difficult to escape. For this purpose, we trigger a recombination operation to create a new starting point for the tabu search procedure, which is explained in the next selection.

2.6   Recombination Operator

The recombination operator aims to transfer good properties of parents to their descendants. The recombination pseudo-code is given in Algorithm 1.

Algorithm 1. The recombination operator

Require: two parent solutions I_1 and I_2
Ensure: An offspring solution I_0
1: $I_0 \leftarrow \emptyset$, $D_1 \leftarrow \emptyset$, $D_2 \leftarrow \emptyset$
2: Sort the bids in each parent according to their prices
3: **while** I_1 and I_2 are not empty **do**
4:     $D_1 \leftarrow first\_element(I_1)$
5:     $D_2 \leftarrow first\_element(I_2)$
6:     if D_1 and/or D_2 are not conflicting with the bids in I_0, add D_1 and/or D_2 to I_0
7:     remove D_1 from I_1
8:     remove D_2 from I_2
9: **end while**
10: Return Child I_0

Given two parent allocations I_1 and I_2 from the elite solution archive which share the highest number of bids, the recombination operator constructs the offspring I_0 in k steps until all the bids of the two parents are visited. Our recombination operator is inspired by the idea of backbone used in [2,22]. In the first step, the set of bids shared by the parents are identified and directly transferred to I_0. Then the following steps are performed:

- Choose the bid with the lowest price from each parent (lines 4 and 5, Algorithm 1).
- The two selected bids are candidate bids that can be inserted in the offspring, if they are not conflicting bids. This is done by conserving the best bids with the highest revenue (lines 6 and 7, Algorithm 1).
- Remove the selected bids from their parents, even if they are not inserted in the offspring (lines 9 and 10, Algorithm 1).
- Repeat the previous steps until all the bids of the parents are examined and removed.

An example of this recombination operation is provided in Fig. 1.

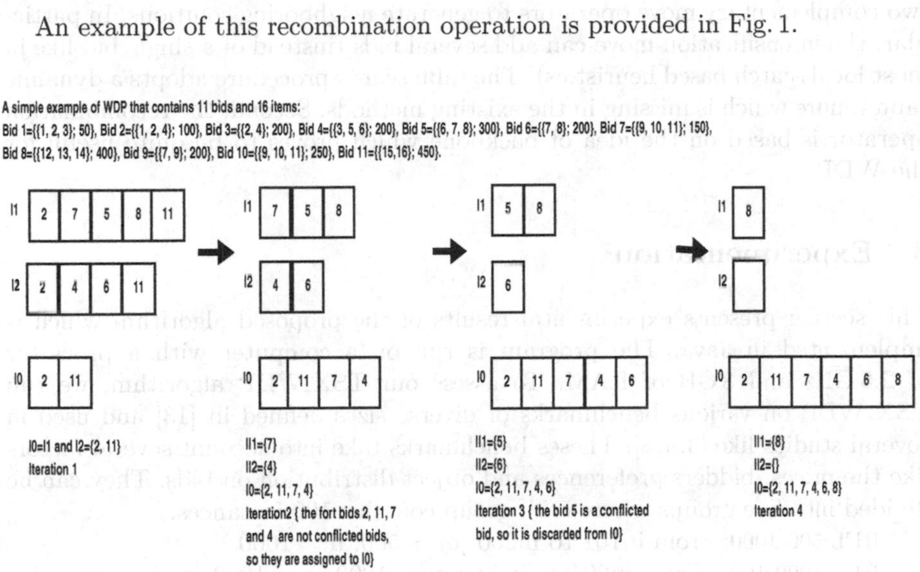

A simple example of WDP that contains 11 bids and 16 items:
Bid 1={{1, 2, 3}; 50}, Bid 2={{1, 2, 4}; 100}, Bid 3={{2, 4}; 200}, Bid 4={{3, 5, 6}; 200}, Bid 5={{6, 7, 8}; 300}, Bid 6={{7, 8}; 200}, Bid 7={{9, 10, 11}; 150}, Bid 8={{12, 13, 14}; 400}, Bid 9={{7, 9}; 200}, Bid 10={{9, 10, 11}; 250}, Bid 11={{15,16}; 450}.

Fig. 1. An example of the recombination operator

2.7   The TSX_WDP Algorithm

The general TSX_WDP algorithm is formalized in Algorithm 2. The algorithm starts with an empty allocation in which no bid is chosen and tries to improve it by looking for a better solution in the current neighborhood. In each iteration, the best authorized bids are selected among the candidate bids to be included in the current allocation. This is achieved with the intensification move (lines 7–9 of Algorithm 2). When no bid can be found to increase the revenue with the intensification move, TSX_WDP switches to the perturbation move by choosing a random bid from the candidate bids (line 11 of Algorithm 2). In both cases, the choice of the bids depends on the status of the tabu list which is updated after each move. Any conflicting bids in the current allocation, when new bids are considered, are removed (lines 13 and 14 of Algorithm 2). The search process is repeated for a fixed number *Itermax* of iterations. During these *Itermax* iterations, if the current best solution cannot be updated for consecutive p (fixed experimentally) moves, the best local optimum found so far is inserted into the archive P and the recombination operator is activated to generate a new starting point for a new round of the tabu search procedure (lines 20–25 of Algorithm 2).

2.8 Discussion

The proposed TSX_WDP algorithm distinguishes itself from the existing heuristic approaches by several features. First, its tabu search procedure is based on two complementary move operators to generate neighboring solutions. In particular, the intensification move can add several bids (instead of a single bid like in most local search based heuristics). The tabu search procedure adopts a dynamic tabu tenure which is missing in the existing methods. Second, the recombination operator is based on the idea of backbone which proves to be quite useful for the WDP.

3 Experimentation

This section presents experimental results of the proposed algorithm which is implemented in Java. The program is run on a computer with a processor of 2.5 GHz and 8 GB of RAM. To assess our TSX_WDP algorithm, we run TSX_WDP on various benchmarks of diverse sizes defined in [13] and used in several studies like [3,4,8]. Theses benchmarks take into account several factors like the prices, bidders preferences and object distribution on bids. They can be divided into five groups where each group contains 100 instances.

-REL 500-1000: From in101 to in200: m = 500, n = 1000
-REL 1000-1000: From in201 to in300: m = 1000, n = 1000
-REL 1000-500: From in401 to in500: m = 1000, n = 500
-REL 1000-1500: From in501 to in600: m = 1000, n = 1500
-REL 1500-1500: From in601 to in700: m = 1500, n = 1500

We calibrated the parameters of the proposed algorithms by an experimental study: The maximum number of iterations (*itermax*) is fixed to 200 and the parameter responsible for the tabu tenure λ is fixed to 0.00006. Each of the 500 instance is solved 40 times independently by the TSX_WDP algorithm with different random seeds.

3.1 Experimental Results

In Table 1, we present the computational results of the TSX_WDP algorithm on the five groups of benchmarks. Given that there are 500 instances, we show

Algorithm 2. TSX_WDP for the Winners Determination Problem

Require: A matrix M, a parameter $Itermax$, Vector of bids B, Parameter p
Ensure: A vector of winning bids A^* and its revenue $f(A^*)$

   $Iter \leftarrow 0$ {Iteration counter}, Initiate $tabu\_list$
   $A^* \leftarrow A \leftarrow \varnothing$
   $opt \leftarrow 0$ {An counter that is incremented if the current solution does not improve in two consecutive iterations; opt returns to 0, when it exceeds the value p, after activating the recombination operator}
   initialize $tabu\_list$
   $P \leftarrow \varnothing$ {Archive of the best local optima encountered A^*}
   while ($Iter < Itermax$) **do**
      Construct neighborhoods from A based on the intensification move
      if There exists an intensification move **then**
         Choose an overall best allowed neighbor A' according to max gain criterion and by considering M {to remove from A' any conflicting bid) {Sect. 2.3}
      else
         Apply the perturbation move {Sect. 2.3} by choosing a random bid from B to create a neighbor A'
      end if
      $A \leftarrow A'$ (Move to the selected neighboring solution A')
      Update $tabu\_list$ {Sect. 2.4} and B {delete the winner bids from B and add the looser bids in it}
      if $f(A) > f(A^*)$ **then**
         $A^* \leftarrow A$
      else
         $opt \leftarrow opt + 1$
      end if
      if $opt = p$ **then**
         Add A^* to the Archive P
         $I_1, I_2 \leftarrow$ Parent_Selection(P) {Sect. 2.5}
         $I_0 \leftarrow$ Recombination_Operator(I_1, I_2) {Sect. 2.6}
         $A \leftarrow I_0$
         $opt \leftarrow 0$
      end if
      $Iter \leftarrow Iter + 1$
   end while
   return (A^* and $f(A^*)$)

only some results of each group, like in some recent papers [4]. For each presented instance, the following computational statistics are indicated: the *maximum revenue* obtained by the TSX_WDP algorithm over the 40 independent trials (Rbest), the *average revenue* over the 40 trials (Ravg), the *worst revenue* over the 40 trials (Rworst) and the average CPU time in seconds (AvgTime). As one can observe, the values of Ravg are very close to the values of Rbest in most of cases and these two values are even equal for certain instances (for example for in101, in102, in205 etc.). This table shows the proposed algorithm can consistently reach high quality solutions for the tested problems.

Table 1. Results obtained by TSX_WDP for WDP benchmarks

Instances	Rbest	Ravg	Rworst	AvgTime	Instance	Rbest	Ravg	Rworst	AvgTime
in101	69585.298	69585.298	69585.298	88	in201	81557.742	80383.277	79331.63	56
in102	72518.222	72518.222	72518.222	76	in202	89289.573	86815.261	81291.193	52
in103	69730.618	69475.485	65903.632	75	in203	86239.213	83941.410	77220.427	54
in104	71327.641	70765.941	65948.396	78	in204	84879.397	84374.869	76822.810	55
in105	73351.044	71570.624	68899.994	93	in205	83748.837	83748.837	83748.837	57
in401	77417.482	77191.182	70628.481	12	in501	83738.040	83506.552	82605.443	107
in402	76273.336	76153.051	74469.073	10	in502	83297.340	82546.590	76751.565	82
in403	74843.958	74356.247	69989.28	10	in503	83718.749	82017.955	78112.719	81
in404	78761.690	78597.224	77939.364	10	in504	83944.901	82772.535	77217.558	76
in405	75915.900	75640.510	74899.125	10	in505	83071.930	81876.413	78909.275	66
in601	107246.248	102862.848	96840.461	117	in602	99668.269	97854.579	91452.904	78
in603	98577.454	96567.287	95219.36	75	in604	101713.602	100786.326	99395.413	78

Table 2. Comparative results of TSX_WDP with Casanova, MA, SLS, TS, SAGII on the WDP benchmarks: μ is the average of the best objective value of the 100 instances in each group. *time* is the average time to reach the best solution.

Test set	100 instances	REL-500-1000	REL-1000-500	REL-1000-1000	REL-1000-1500	REL-1500-1500
TSX_WDP	Time	74.19	9.45	48.98	75.92	90.61
	μ	**69647.975**	**75274.184**	**86786.159**	**85577.806**	**103178.732**
Casanova	Time	119.46	57.74	111.42	168.24	165.92
	μ	37053.78	51248.79	51990.91	56406.74	65661.03
$\delta_{TSX/Casanova}(\%)$		46.79	31.91	40.09	34.08	36.36
TS	Time	91.07	25.84	104.30	223.37	175.68
	μ	65286.94	71985.34	81633.63	77931.41	97824.64
$\delta_{TSX/TS}(\%)$		6.26	4.36	5.93	8.93	5.18
SLS	Time	22.35	5.91	14.19	14.97	16.47
	μ	64216.14	72206.07	82120.31	79065.08	98877.07
$\delta_{TSX/SLS}(\%)$		7.79	4.07	5.37	7.61	4.16
MA	Time	56.64	14.98	33.05	24.51	28.22
	μ	65740.25	73604.62	83304.20	79644.64	99957.96
$\delta_{TSX/AM}(\%)$		5.61	2.21	4.01	6.93	3.12
SAGII	Time	38.06	24.46	45.37	68.82	91.78
	μ	64922.02	73922.10	83728.34	82651.49	101739.64
$\delta_{TSX/SAGII}(\%)$		6.78	1.79	3.52	3.41	1.39

3.2   Comparative Results for the WDP

In order to further show the effectiveness of the TSX_WDP algorithm, we present a comparative study with five state of the art algorithms from the literature: Casanova [10], SAGII [8], SLS [3], TS [3], MA [4].

In Table 2, we show the general comparative results for each group. In this table, rows μ correspond to the average of best objective value of the 100 instances in each group. Rows *time* represent the average time to reach the best solution. $\delta(\%)$ is the deviation of the TSX_WDP algorithm with respect to each reference algorithm. The deviations are calculated respectively as follows: $\mu_{TSX\_WDP} - \mu_{algo\_X})/\mu_{TSX\_WDP}$ where *algo_X* is one of the five reference algorithms. Since the compared algorithms are implemented in different languages and run on different platforms, the comparison is focused on solution quality that can be reached by each algorithm. The computing time is provided only for indicative purposes. The results of the reference algorithms are extracted from the corresponding papers except the results of Casanova which are from [8].

Table 2 discloses that TSX_WDP gives an improvement between 31 % and 47 % in solution quality compared to Casanova. TSX_WDP finds better solutions with shorter times than Casanova. TSX_WDP shows good performances compared to SLS. The improvement is between 4 % and 8 %. The results of TSX_WDP are better than TS in quality and in time (with an improvement rate between 4 % and 9 %). TSX_WDP outperforms MA. The deviation is between 2 % and 7 %. Finally, TSX_WDP produces better results than SAGII which is currently the most successful algorithm for the WDP and is based on sophisticated Branch-and-Bound and preprocessing tools (The deviation is between 1 % and 7 %). Thus, we can conclude that TSX_WDP discovers new best results for the five groups of benchmarks.

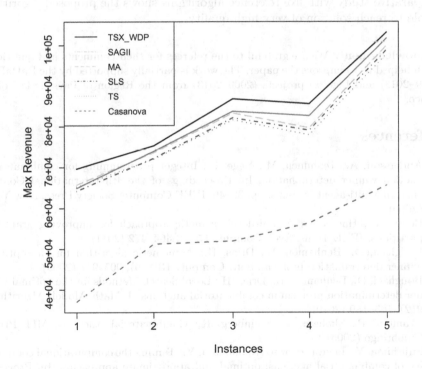

Fig. 2. A comparison of the solution quality between TSX_WDP, Casanova, TS, SLS, MA and SAGII

To further illustrate the results of Table 2, we consider the comparative curves of Fig. 2. The X-axis of the curves represent the 5 groups of the WDP benchmarks and their Y-axis are the gain of each group (μ). These curves confirm that TSX_WDP competes favorably with each of the reference algorithms for each group of instances.

4   Conclusion

In this work, we have presented a tabu search algorithm for the winner determination problem based on a two different neighborhood structures and a recombination operator. The algorithm uses the intensification move to improve progressively the quality of the current solution. When the solution cannot be further improved, the TSX_WDP algorithm switches to a perturbation move by choosing a random bid. In both cases, a tabu list is used to prevent the search from revisiting the previous examined solutions. To escape deep local optima, the proposed algorithm employs a backbone-based recombination operator which relies on an elite solution archive which is built and updated during the search. This recombination operator generates new starting points for tabu search with the aim of leading the algorithm into new promising search areas. The proposed TSX_WDP algorithm is evaluated on a set of 500 benchmark instances. The comparative study with five reference algorithms shows the proposed algorithm is able to reach solution of very high quality.

Acknowledgment. We are grateful to the referees for their comments and questions which helped us to improve the paper. The work is partially supported by the RaDaPop (2009–2013) and LigeRO projects (2009–2013) from the Region of Pays de la Loire, France.

References

1. Andersson, A., Tenhunen, M., Ygge, F.: Integer programming for combinatorial auction winner determination. In: Proceedings of the 4th International Conference on Multi-agent Systems, pp. 39–46. IEEE Computer Society Press, New York (2000)
2. Benlic, U., Hao, J.K.: A multilevel memetic approach for improving graph k-partitions. IEEE Trans. Evol. Comput. **15**(5), 464–472 (2011)
3. Boughaci, D., Benhamou, B., Drias, H.: A memetic algorithm for the optimal winner determination problem. Soft. Comput. **13**(8–9), 905–917 (2009)
4. Boughaci, D., Benhamou, B., Drias, H.: Local Search Methods for the optimal winner determination problem in combinatorial auctions. J. Math. Model. Algorithms **9**(2), 165–180 (2010)
5. Cramton, P., Shoham, Y., Steinberg, R.: Combinatorial Auctions. MIT Press, Cambridge (2006)
6. Fujishima, Y., Leyton-Brown, K., Shoham, Y.: Taming the computational complexity of combinatorial auctions: optimal and approximate approaches. In: Proceedings of the 16th International Joint Conference on Artificial Intelligence, Sweden, pp. 48–53 (1999)

7. Galinier, P., Hao, J.K.: Hybrid evolutionary algorithms for graph coloring. J. Comb. Optim. **3**(4), 379–397 (1999)
8. Guo, Y., Lim, A., Rodrigues, B., Zhu, Y.: Heuristics for a bidding problem. Comput. Oper. Res. **33**(8), 2179–2188 (2006)
9. Holland, A., O'sullivan, B.: Towards fast Vickrey pricing using constraint programming. Artif. Intell. Rev. **21**(3–4), 335–352 (2004)
10. Hoos, H.H., Boutilier, C.: Solving combinatorial auctions using stochastic local search. In: Proceedings of the 17th National Conference on Artificial Intelligence, USA, pp. 22–29 (2000)
11. Abrache, J., Crainic, T.G., Gendreau, M., Rekik, M.: Combinatorial auctions. Ann. Oper. Res. **153**(1), 131–164 (2007)
12. Klemperer, P.: Auctions: Theory and Practice. Princeton University Press, Princeton (2004)
13. Lau, H.C., Goh, Y.G.: An intelligent brokering system to support multi-agent web-based 4th-party logistics. In: Proceedings of the 14th International Conference on Tools with Artificial Intelligence, Washington, DC, pp. 54–61 (2002)
14. Lehmann, D., Rudolf, M., Sandholm, T.: The winner determination problem. In: Cramton, P., et al. (eds.) Combinatorial Auctions. MIT Press, Cambridge (2006)
15. Leyton-Brown, K., Shoham, Y., Tennenholtz, M.: An algorithm for multi-unit combinatorial auctions. In: Proceedings of the 17th National Conference on Artificial Intelligence and 12th Conference on Innovative Applications of Artificial Intelligence, Austin, Texas, pp. 56–61. AAAI Press/MIT Press (2000)
16. Nisan, N.: Bidding and allocation in combinatorial auctions. In: Proceedings of the 2nd ACM Conference on Electronic Commerce, pp. 1–12. ACM Press, Minneapolis (2000)
17. Rothkopf, M.H., Pekee, A., Ronald, M.: Computationally manageable combinatorial auctions. Manage. Sci. **44**(8), 1131–1147 (1998)
18. Sandholm, T.: Optimal winner determination algorithms. In: Cramton, P., et al. (eds.) Combinatorial Auctions. MIT Press, Cambridge (2006)
19. Sandholm, T., Suri, S.: Improved optimal algorithm for combinatorial auctions and generalizations. In: Proceedings of the 17th National Conference on Artificial Intelligence, USA, pp. 90–97 (2000)
20. Sandholm, T., Suri, S., Gilpin, A., Levine, D.: CABoB: a fast optimal algorithm for combinatorial auctions. In: Proceedings of the International Joint Conferences on Artificial Intelligence, Seattle, WA, pp. 1102–1108 (2001)
21. Vries, S., Vohra, R.: Combinatorial auctions a survey. INFORMS J. Comput. **15**, 284–309 (2003)
22. Wang, Y., Lu, Z., Glover, F., Hao, J.K.: Backbone guided tabu search for solving the UBQP problem. J. Heuristics **19**(4), 679–695 (2013)

Improving Efficiency of Metaheuristics for Cellular Automaton Inverse Problem

Fazia Aboud[✉], Nathalie Grangeon, and Sylvie Norre

LIMOS CNRS UMR 6158, Antenne IUT dAllier,
Avenue Aristide Briand, 03100 Montluçon, France
{aiboud,grangeon,norre}@moniut.univ-bpclermont.fr

Abstract. The aim of this paper concerns several propositions to improve previous works based on a combination between metaheuristic and cellular automaton for the generation of 2D shapes. These improvements concern both the reduction of the search space and of the computational time. The first proposition concerns a new approach which delegates the determination of the number of generation to the cellular automaton. The second proposition consists in the reduction of the number of times the cellular automaton is requested. The last proposition concerns the adaptation of the method by exploiting the properties of the expected shape, in particular in case of symmetric shapes. Obtained results show that these propositions permit to improve the results as well as the computational times and the quality of the solution.

1 Introduction

Cellular automata were introduced by Stanislas Ulam and John Von Neumann in an attempt to model natural physical and biological systems [1,2]. They are discrete dynamic systems in space and time with simple local interactions but complex global behavior [3]. They can be used to study complex dynamic systems such as self-organization phenomena [1], development of tumor [4], fire forest propagation [5], diffusion phenomena [6] and shapes generation [7].

Cellular automaton consists in a regular lattice of cells. The communication between cells is limited to local interaction. Each cell can take a state chosen among a finite set of states. This state can evolve over time depending on the states of its neighbor determined by the interaction system through a local evolutionary rule. The set of these local rules forms a transition function of the cellular automaton.

In this paper, we consider the combination of cellular automaton and simulated annealing to generate 2D binary preset shapes, also called inverse problem. This work comes within the scope of preliminary studies of simulation of morphogenesis process with the objective to simulate, by cellular automata, the generation of full organ from a single cell. The morphogenesis is an important process which allows living things to develop organized structures thanks to the interactions of cells either among themselves or with their environment.

© Springer International Publishing Switzerland 2014
P. Legrand et al. (Eds.): EA 2013, LNCS 8752, pp. 168–179, 2014.
DOI: 10.1007/978-3-319-11683-9_14

In the next part, we present this inverse problem and a first approach to solve it. The objective of this paper is to propose ways to address the drawbacks of this first approach: to reduce the size of the solution, to accelerate the computational time and to take into account the properties of the expected shape. These points are presented in the third part. The fourth part gives numerical results.

2   Context

2.1   Cellular Automaton Inverse Problem

In order to obtain a given configuration of cellular automaton or behavior after a given number of generations, it is very important to determine the initial configuration and/or the transition function allowing this configuration or this behavior. This problem is known as the inverse problem. The inverse problem of deducing the local rules from a given configuration or global behavior is extremely hard [8].

Several works in the literature propose to use the evolutionary computation techniques such as genetic algorithm to solve this inverse problem. These problems concern the computational tasks (density classification and synchronization) [9,10] and the generation of full shapes [2,7].

In this paper, we consider the generation of any shapes: full shapes and hollow shapes (shapes with holes). The generation of hollow shapes (shapes with holes) is really important in order to consider biological phenomena where living cells may die.

The generation of shapes by cellular automaton is a particular case of the inverse problem of the cellular automaton. Two problems arise: the first one is how to choose a transition function and the second one is how often to apply this transition function (number of generations) allowing to the cellular automaton to evolve from an input configuration (initial shape) toward output configuration (expected shape).

The Fig. 1 illustrates the studied problem: how to obtain a shape Sh* from an initial shape Sh0 with a cellular automaton? In other words, the objective is to determine a transition function F* and a number of generation Ng^* which maximize a similarity criterion $C(Sh, Sh^*)$:

$$\text{determine } F^* \text{ and } Ng^* \text{ such as } C(Sh_{F^*,Ng^*}, Sh^*) = \max_{F,Ng}(C(Sh_{F,Ng}, Sh^*))$$

$$(1)$$

where $Sh_{F,Ng}$ is the shape obtained after applying Ng times the transition function F.

2.2   A First Approach

A first approach has been presented in [15] and consists in the combination of simulated annealing and cellular automaton. This approach uses a binary solution encoding composed of two parts: one for the transition function F and one

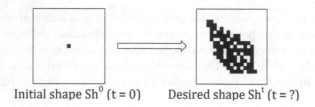

Initial shape Sh⁰ (t = 0)   Desired shape Shᵗ (t = ?)

Fig. 1. Inverse problem

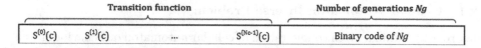

Fig. 2. Solution encoding

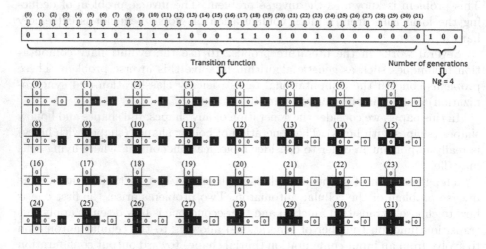

Fig. 3. An example of solution X

for the number of generations Ng. This encoding uses the notion of combination of states (states of the cells of the interaction system) and is represented by Fig. 2 where $S^{(i)}(c)$ represents the state of cell c at time $t + 1$ according to the combination (i) at time t. $Nc = k^n$ is the number of combination of states in the interaction system where n is the size of the interaction system and k is the number of states. The size of the interaction system is the number of considered neighbor cells.

Figure 3 shows an example of a solution with Von Neumann interaction system with five neighbor cells ($n = 5$):

– the transition function describes the local rule applied to each combination of the interaction system. The sixteen first combinations are the local evolutionary rules when the observed cell is empty. The sixteen next combinations describe the local evolutionary rules when the observed cell is occupied.

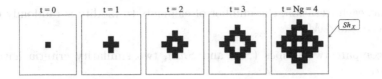

Fig. 4. Evolution of the cellular automaton over times

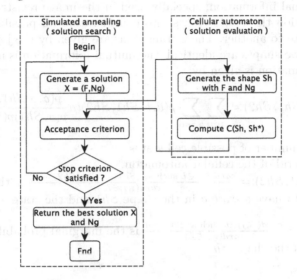

Fig. 5. First approach (named (F, Ng))

The combination number is given in parenthesis. For each combination, the state of the cell and its neighbor cells is detailed.

– the number of generations Ng is equal to 4.

Figure 4 shows the evolution of the cellular automaton over time by using the solution given in Fig. 3. The last shape will be compared to the desired shape.

The simulated annealing starts with an initial solution randomly generated. With this solution, the cellular automaton generates a shape $Sh_{F,ng}$ that is compared to the expected shape Sh^*. At each iteration, the simulated annealing chooses randomly a new solution in the neighborhood system of the current solution. The cellular automaton then generates a new shape and the new solution is accepted according to an acceptance criterion. The principle is depicted by Fig. 5.

Three neighborhood systems have been proposed depending on the number of modified elements. The modification of an element consists in choosing randomly and uniformly a new state among the possible states.

– $\Gamma 1$ a single element is modified,
– $\Gamma 2$ two elements are modified,

– ΓM at each iteration, the number of elements to modify is randomly chosen between 1 and M.

To compare two shapes ($Sh1$ and $Sh2$), two similarity criteria have been proposed:

– ID the number of identical cells in the two shapes,
– MI the mutual information, specially used in the image registration domain and information theory [11] takes into consideration all possible transitions from one state to another. These transitions are given by the joint probabilities. When two shapes are identical, the mutual information is maximal. The mutual information is given by:

$$MI(Sh1, Sh2) = \sum_{a\in\alpha} \sum_{b\ ina} p(a, b, Sh1, Sh2) ln \frac{p(a, b, Sh1, Sh2)}{p(a, Sh1)p(b, Sh2)} \qquad (2)$$

- α is the number of possible cell states,
- L is the grid of the cellular automaton,
- $p(a, b, Sh1, Sh2) = \frac{card(i,j)\in L \text{ such as } (Sh1_{i,j}=a)\wedge(Sh2_{i,j}=b)}{card(i,j)\in L}$ is the joint probabilities to have a state a in the shape $Sh1$ and the state b in the shape $Sh2$.
- $p(a, Sh) = \frac{card((i,j)\in L \text{ such as } Sh_{i,j}=a}{card(i,j)\in L)}$ is the marginal probability to have a state a in the shape Sh.

2.3   Conclusion

The size of the solutions space Ω is the product of the number of possible transition functions by the maximum number of iterations $Ngmax$ expressed in the solution encoding (size of the second part).

$$|\Omega| = k^{Nc} \times Ngmax \qquad (3)$$

The number of possible transition functions k^{Nc} depends on the size of the transition function Nc which depends on the size of the interaction system and the number of possible states of cell. When we use a cellular automaton with two states and Moore interaction system composed of 8 neighbor cells and the observed cell, the size of the solution is 512 and the number of possible transition functions is 2^{512}.

Table 1 presents the size of the solution space according to the number of states of the cellular automaton k, the considered interaction system (Moore or Von Neumann) and the maximum number of generation coded in the solution.

The size of search space increases exponentially with the number of states (k) and $Ngmax$.

Obtained results are promising, but some improvement should be done particularly in terms of computation time and search space.

Table 1. Size of solution space

$Ngmax$	Von Neumann			Moore		
	3	7	15	3	7	15
k	2	1,72E+10	3,44E+10	6,87E+10	6,36E+154	1,07E+155	2,15E+155
	3	7,78E+116	2,35E+117	7,06E+117	∞	∞	∞
	4	∞	∞	∞	∞	∞	∞

3 Proposition

3.1 A New Approach to Reduce the Size of Solution

A possible way to reduce the search space is to delegate the determination of Ng to the cellular automaton. To do so, we propose a solution encoding based only on the transition function.

Figure 6 gives the principle of the combination. At each iteration, the metaheuristic provides to the cellular automaton a solution $X = (F)$ and the cellular automaton builds different shapes by applying $Ngmax$ times this transition function F to the initial shape. At each generation $i \in [1, Ngmax]$, the obtained shape $Sh_{(F,i)}$ is compared to the expected shape $Sh*$. The shape with the best similarity criterion $C(Sh_{X,Ng}, Sh*)$ is recorded:

$$\text{such as } C(Sh_{F,Ng}, Sh*) = \max_{i=1, Ngmax}(C(Sh_{F,i}, Sh*)) \tag{4}$$

The cellular automaton returns to the metaheuristic $C(Sh_{X,Ng}, Sh*)$ and the corresponding number of generations Ng.

With this approach, the size of the solutions space Ω is equal to the number of possible transition functions ($|\Omega| = k^{Nc}$).

3.2 To Reduce the Computational Time

In the first approach, at each iteration of the metaheuristic, the cellular automaton applies the transition function Ng times whereas in the proposed approach, the transition function is applied $Ngmax$ times. The proposition reduces the search space but may increase the computational time.

To face this drawback, we propose to reduce the number of times the cellular automaton is requested. In fact, during an iteration of the simulated annealing, the transition is applied $Ngmax$ times and during these iterations some combinations are never used. Instead of requesting the cellular automaton at each iteration of the simulated annealing, we propose to use the cellular automaton only if the modification proposed by the simulated annealing can have a consequence on the similarity criterion. For that, when the transition function is applied, the used and unused combinations are identified. If a combination is not used and if this combination is selected during the next iteration of the simulated annealing algorithm, the modification is accepted without using the

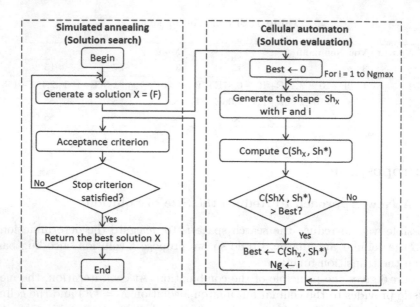

Fig. 6. Proposed approach

cellular automaton as it is sure that the similarity criterion is not modified. As will be seen in the next part, such a modification permits to reduce significantly the computational time (with a ratio from 3 to 6).

3.3   To Take into Account the Properties of the Expected Shape

Another way to improve the results by reducing the search space is to adapt the proposed method by exploiting the properties of the expected shape. It is the case for symmetric shapes. In the literature, some works exploit the symmetry:

– In [12] the different types of symmetry in the spatiotemporal diagram of one dimensional cellular automaton are presented. In these works, the authors study the symmetric cellular automaton, but they have not interested to the inverse problem.
– Reference [13] is interested to seek cellular automata that perform universal computational tasks. The only binary automaton currently identified as supporting universal computation is "game of life". Its ability to simulate a Turing machine is proved, using gliders (periodic patterns which, when evolving alone, are re-produced identically after some shift in space), glider guns, and eaters. The glider gun emits a glider stream that carries information and creates logic gates through collisions. The eaters permit, by absorbing gliders, the creation of logic circuits using any combination of logic gates. A transition function, which exploits symmetry, is used to search glider which can displaced in the space in all direction.

We have identified different symmetries (Fig. 7).

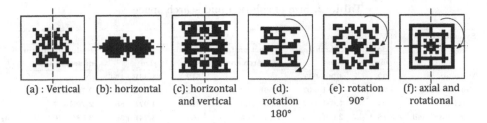

(a) : Vertical   (b): horizontal   (c): horizontal   (d):   (e): rotation   (f): axial and
                                      and vertical   rotation   90°   rotational
                                                      180°

Fig. 7. Type of symmetry

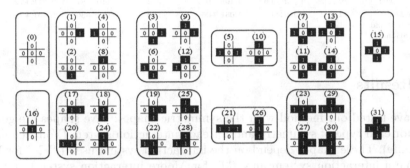

Fig. 8. Equivalent combinations of the interaction system state by horizontal, vertical and rotational symmetry

In order to reduce the search space of the metaheuristic, we propose a transition function which takes into account this symmetry. The symmetry in cellular automaton may be defined by the invariance of the evolutionary local rules according to a spatial transformation T of the interaction system state. The spatial transformation T of the interaction system is a permutation of the order of state of cells in the interaction system.

To exploit this symmetry, the combinations of the interaction system state which are symmetric at the generation t should have the same state at the generation $t + 1$. Then, an evolutionary local rule is applied to a set of symmetric combinations. These evolutionary local rules form the transition function of the cellular automaton.

For example, in case of horizontal, vertical and rotational symmetry, the transition function in case of Von Neumann interaction system is reduced to 12 combinations instead of 32 combinations. The corresponding equivalent combinations of the interaction system state are given in Fig. 8.

Table 2 presents the effects of the modifications of the solution encoding on its size and on the size of the search space according to the considered interaction system (Moore or Von Neumann). We can see that the size of the search space is largely reduced whatever the symmetry. In case of Von Neumann interaction system, the reduction is so important that an enumeration of all solutions in the search space is possible.

Table 2. Size of solution and search space

Symmetry	Von Neumann			Moore		
	Solution size	Search space size	Reduction factor of search space	Solution size	Search space size	Reduction factor of search space
Without	32	4.293+09	-	512	1.34E+154	
Horizontal	24	1.68E+07	2.56E+02	288	4.97E+86	2.70E+67
Vertical	24	1.68E+07	2.56E+02	288	4.97E+86	2.70E+67
Horizontal and vertical	18	2.62E+05	1.64E+04	168	3.74E+50	3.58E+103
Rotation with 90	12	4.10E+03	1.05E+06	140	1.39E+42	9.62E+111
Rotation with 180	20	1.05E+06	4.10E+03	272	7.59E+81	1.77E+72
Horizontal, vertical and rotational	12	4.10E+03	1.05E+06	102	5.07E+30	2.64E+123

4 Results

We have tested our method with the symmetric shapes. Five shapes are generated for each type of symmetry and for each interaction system (Von Neumann or Moore). These shapes depend on the interaction system used ("VN" for Von Neumann interaction system and "M" for Moore interaction system) and the type of symmetry ("H" for Horizontal symmetry, "V" for Vertical symmetry, "HV" for the Horizontal and Vertical symmetry, "HVR" for Horizontal, Vertical and Rotational symmetry, "R90-" for the Rotation by 90 and "R180-" for the Rotation by 180). With Von Neumann interaction system, R90 and HVR are equivalent so no shapes have been generated with R90.

For each shape $Sh*$, we run the proposed algorithm by using either the mutual information (MI) or the number of identical cells (ID) as similarity criterion. For each shape generated with Von Neumann interaction system, we have considered the two interaction systems: Von Neumann and Moore. For each shape generated with Moore, we have only considered Moore interaction system. As Moore interaction system is an extension of Von Neumann system, it seems not suitable to consider Von Neumann interaction system to obtain shapes generated with Moore interaction system.

The parameters of the simulated annealing are: number of iterations $NIter = 5.10^6$ iterations, initial temperature $T0 = 10$ and final temperature $Tf = 10^{-4}$. As the simulated annealing is a stochastic algorithm, five replications are done in each case. For the neighborhood system ΓM, M is fixed to 10.

To compare our algorithms, we compute for each shape and each replication the ratio R between the overlapping cells of the obtained shape and the expected shape and, the total number of cells of the expected shape. This ratio is given by:

$$R(Sh, Sh*) = \frac{ID(Sh, Sh*)}{ID(Sh*, Sh*)} \tag{5}$$

Results are given in Table 3. With Von Neumann interaction system, the optimal solution is obtained at each replication for all shapes generated with Von

Table 3. Results obtained with symmetric shapes

			Shapes generated with Von Neumann								Shapes generated with Moore							
			MI				ID				MI				ID			
			[90%,95%[	[95%,98%[	[98%,100%[	100%	[90%,95%[	[95%,98%[	[98%,100%[	100%	[90%,95%[	[95%,98%[	[98%,100%[	100%	[90%,95%[	[95%,98%[	[98%,100%[	100%
H	Without symmetry	Γ1	0.12	0.32	0.4	0.16	0	0	0.92	0.08	0.16	0.16	0.64	0	0	0.4	0.44	0.16
		Γ2	0.08	0.32	0.44	0.16	0	0.08	0.8	0.12	0.16	0.2	0.56	0.04	0	0.28	0.72	0
		ΓM	0	0.28	0.52	0.2	0	0	0.84	0.16	0.16	0.24	0.56	0.04	0.04	0.2	0.68	0.08
	With symmetry	Γ1	0	0	0.68	0.32	0	0	0.56	0.44	0.04	0.32	0.28	0.36	0	0.28	0.4	0.32
		Γ2	0	0	0.44	0.56	0	0	0.68	0.32	0	0.28	0.32	0.4	0	0.2	0.4	0.4
		ΓM	0	0	0.64	0.36	0	0	0.48	0.52	0	0.16	0.52	0.32	0	0.32	0.32	0.36
V	Without symmetry	Γ1	0.12	0.32	0.4	0.16	0	0	0.68	0.32	0.16	0.16	0.64	0	0	0.08	0.72	0.2
		Γ2	0.08	0.2	0.48	0.24	0	0.12	0.48	0.4	0.04	0.24	0.56	0.16	0	0	0.8	0.2
		ΓM	0.08	0.28	0.4	0.24	0	0.08	0.52	0.4	0	0.16	0.76	0.08	0	0.04	0.84	0.12
	With symmetry	Γ1	0	0	0.68	0.32	0	0	0.2	0.8	0.04	0.32	0.28	0.36	0	0	0.28	0.72
		Γ2	0	0.04	0.32	0.64	0	0.04	0.28	0.68	0	0	0.6	0.4	0	0	0.2	0.8
		ΓM	0	0	0.28	0.72	0	0	0.28	0.72	0	0	0.52	0.48	0	0	0.16	0.84
HV	Without symmetry	Γ1	0	0.32	0.56	0.12	0.16	0.2	0.52	0.12	0.16	0.44	0.2	0.04	0.16	0.04	0.6	0.2
		Γ2	0	0.28	0.44	0.28	0.12	0.12	0.56	0.2	0.28	0.2	0.32	0.04	0.16	0.16	0.32	0.36
		ΓM	0	0.32	0.28	0.4	0.08	0	0.52	0.4	0.08	0.28	0.48	0	0.08	0.28	0.48	0.16
	With symmetry	Γ1	0	0	0.12	0.88	0	0	0	1	0	0.16	0.12	0.72	0	0.2	0.04	0.76
		Γ2	0	0	0.04	0.96	0	0	0	1	0.04	0	0.2	0.76	0	0.12	0.12	0.76
		ΓM	0	0	0.04	0.96	0	0	0.04	0.96	0.04	0.08	0.08	0.8	0	0.16	0.04	0.8
HVR	Without symmetry	Γ1	0.04	0.28	0.68	0	0	0	0.92	0.08	0.08	0.16	0.36	0.08	0.24	0.36	0.36	0.04
		Γ2	0	0.24	0.68	0.08	0	0	0.88	0.12	0.16	0.36	0.28	0.08	0.32	0.32	0.32	0.04
		ΓM	0	0.12	0.88	0	0.04	0	0.8	0.16	0.16	0.24	0.4	0	0.16	0.24	0.56	0.04
	With symmetry	Γ1	0	0	0	1	0	0	0	1	0	0	0	1	0	0	0	1
		Γ2	0	0	0	1	0	0	0	1	0	0	0	1	0	0	0	1
		ΓM	0	0	0	1	0	0	0	1	0	0	0	1	0	0	0	1
Rot-180°	Without symmetry	Γ1	0	0.12	0.44	0.44	0	0.04	0.2	0.76	0.08	0.6	0.32	0	0.04	0.16	0.68	0.12
		Γ2	0	0.08	0.28	0.64	0	0	0.12	0.88	0	0.44	0.52	0.04	0	0.2	0.72	0.08
		ΓM	0	0.12	0.28	0.6	0	0	0.24	0.76	0	0.28	0.56	0.16	0.04	0.08	0.68	0.2
	With symmetry	Γ1	0	0	0.16	0.84	0	0	0.16	0.84	0	0.2	0.52	0.28	0	0.16	0.48	0.36
		Γ2	0	0	0.12	0.88	0	0	0.08	0.92	0	0.12	0.32	0.56	0	0.12	0.36	0.52
		ΓM	0	0	0.16	0.84	0	0	0.16	0.84	0.04	0.08	0.44	0.44	0	0.04	0.4	0.56
Rot-90°	Without symmetry	Γ1	-	-	-	-	-	-	-	-	0.44	0.32	0.16	0	0.36	0.16	0.4	0.08
		Γ2	-	-	-	-	-	-	-	-	0.4	0.28	0.2	0	0.24	0.32	0.44	0
		ΓM	-	-	-	-	-	-	-	-	0.28	0.4	0.24	0	0.24	0.32	0.36	0.08
	With symmetry	Γ1	-	-	-	-	-	-	-	-	0	0.16	0.28	0.56	0	0.04	0.24	0.72
		Γ2	-	-	-	-	-	-	-	-	0	0.08	0.24	0.68	0	0.12	0.2	0.68
		ΓM	-	-	-	-	-	-	-	-	0	0.04	0.28	0.68	0	0.08	0.08	0.84

Neumann, taking into account or not the symmetry. The computational time of the proposed method for symmetric shape is of the order of second while it is of order of minute when the symmetry is not considered. These computational times are of course lower than those of the enumerative method.

The results obtained by using Moore interaction system are presented in Table 3. For each symmetry, each kind of shapes (generated by Von Neumann or by Moore) and each similarity criterion, the frequency distribution of R is given. When the symmetry is considered, R is always superior to 90 % and very often superior to 95 % (in all cases for ID). So we have chosen to consider the classes

[90 %, 95 %[, [95 %, 98 %[and [98 %, 100 %[. The column 100 % is added in order to identify cases where the expected shape is obtained (optimal solution).

In all cases, the results are improved by taking into account the symmetry. The number of identical cells as performance criterion seems to perform better than mutual information but it is not obvious. Similarly, results obtained by the neighborhood systems are similar and it is not easy not conclude.

With the reduction of search space, we have improved the run time with a factor variable between 2 to 1512 for the shapes generated with Von Neumann interaction system when we use Von Neumann interaction system. With Moore interaction system, we have accelerated our algorithm with factor from 1.2 to 178. The run time depends on the similarity criterion and the type of symmetry.

5   Conclusion

In previous works, we have proposed a combination of simulated annealing and binary cellular automata to generate 2D shapes. The principle was the following: at each iteration, the simulated annealing chooses randomly a new solution (a solution was composed of a transition function and a number of generations) in the neighborhood system of the current solution. The cellular automaton then generated a new shape and the new solution was accepted according to an acceptance criterion. Obtained results were promising but some improvements appeared to be necessary.

This is the aim of this paper in which different propositions are given in order to reduce the computational time and/or the search space. The first improvement concerns the proposition of a new approach which delegates the determination of the number of generation to the cellular automaton. The second proposition consists in the reduction of the number of times the cellular automaton is requested. The last proposition concerns the adaptation of the method by exploiting the properties of the expected shape, in particular in case of symmetric shapes. Obtained results show that these propositions permit to improve the results as well as the computational times and the quality of the solution.

Our future works consist to test other neighborhood systems and other objective functions, in order to improve our results. We envisage applying this method to generate complex shapes: from any initial shape, with a number of states greater than 2, 3D shapes. Finally, we envisage adapting our method to generate a set of target shapes and to solve this problem as dynamic optimization problem.

References

1. Kenneth, O., Stanly, K., Miikkulainen, R.: A Taxonomy for Artificial Embryogeny. Artif. Life J. **9**(2), 93–130 (2003)
2. De Garis, H.D.: Artificial embryology and cellular differentiation. In: Bentley, P.J. (ed.) Evolutionary Design by Computers. Morgan Kaufmann Publishers, San Francisco (1999)

3. Wolfram, S.: Universality and complexity in cellular automata. Physica **D10**, 1–35 (1984). North Holland
4. Moreira, J., Deutsch, A.: Cellular automaton models of tumor development: a critical review. Adv. Complex Syst. **5**, 247–267 (2002)
5. Karalyllidis, I.: Acceleration of cellular automata algorithms using genetic algorithms. Adv. Eng. Softw. **30**, 419–437 (1999)
6. Margolus, N.: Physics-like models of computation. Physica **10 D**, 81–95 (1984)
7. Chavoya, A., Duthen, Y.: Using a genetic algorithm to evolve cellular automata for 2D/3D computational development. In: 8th Annual Conference on Genetic and Evolutionary Computation, USA (2006)
8. Ganguly, N., Sikdar, K.B., Deutsch, A., Canright, G., Chaudhuri, P.: A survey on cellular automata. Technical Report, Centre for High Performance Computing, Dresden University of Technology, December 2003
9. Mitchell, M., Crutchfield, J.P., Das, R.: Evolving cellular automata with genetic algorithms: a review of recent work. In: 1st International Conference on Evolutionary Computation and its Applications, Russia (1996)
10. Bäck, T., Breukelaar, R.: Using genetic algorithms to evolve behavior in cellular automata. In: Calude, C.S., Dinneen, M.J., Păun, G., Jesús Pérez-Jímenez, M., Rozenberg, G. (eds.) UC 2005. LNCS, vol. 3699, pp. 1–10. Springer, Heidelberg (2005)
11. Gray, R.: Entropy and Information Theory. Springer, New York (1990)
12. Mainzer, K.: Symmetry and complexity in dynamical systems. Eur. Rev. **13**(2), 29–48 (2005)
13. Sapin, E., Bailleux, O., Jean-Jacques, C.: Research of a cellular automaton simulating logic gates by evolutionary algorithms. In: Ryan, C., Soule, T., Keijzer, M., Tsang, E.P.K., Poli, R., Costa, E. (eds.) EuroGP 2003. LNCS, vol. 2610, pp. 414–423. Springer, Heidelberg (2003)
14. Aiboud, F., Grangeon, N., Norre, S.: Simulated annealing based metaheuristics and binary cellular automata to generate 2D shapes. In: 10th International Conference on Aritificial Evolution, Angers (2011)
15. Aiboud, F., Grangeon, N., Norre, S.: Generation of 2D shapes with cellular automaton and genetic algorithm: extension of Chavoya and Duthen work. In: 9th Metaheuristics International Conference MIC, Italy (2011)

Memetic Algorithms

Memetic Algorithm with an Efficient Split Procedure for the Team Orienteering Problem with Time Windows

Rym Nesrine Guibadj[(✉)] and Aziz Moukrim

Laboratoire Heudiasyc, UMR 7253 CNRS,
Université de Technologie de Compiègne, 60205 Compiègne, France
{rym-nesrine.guibadj,aziz.moukrim}@hds.utc.fr

Abstract. The Team Orienteering Problem (TOP) is a variant of the vehicle routing problem. Given a set of vertices, each one associated with a score, the goal of TOP is to maximize the sum of the scores collected by a fixed number of vehicles within a certain prescribed time limit. More particularly, the Team Orienteering Problem with Time Windows (TOPTW) imposes the period of time of customer availability as a constraint to assimilate the real world situations. In this paper, we present a memetic algorithm for TOPTW based on the application of split strategy to evaluate an individual. The effectiveness of the proposed MA is shown by many experiments conducted on benchmark problem instances available in the literature. The computational results indicate that the proposed algorithm competes with the heuristic approaches present in the literature and improves best known solutions in 101 instances.

1 Introduction

The Orienteering Problem (OP) was firstly introduced by Tsiligirides [24]. The roots of this problem trace back to the pioneering work of Golden et al. [7] who proved that the OP is NP-hard and used it to formulate and solve the home fuel delivery problem. The name "Orienteering Problem" originates from the sport game of orienteering described in [3]. Later, a new variant of the problem called Team Orienteering Problem (TOP) was introduced since it is widely seen in many real life situations, like for example the routing of technicians [21] and fuel delivery problems [7]. Many heuristics have been successfully applied to TOP. There are four methods that can be considered as the state-of-the-art algorithms in the literature: a variable neighborhood search proposed by Archetti et al. [1], a memetic algorithm [2], a path relinking approach [20] and a PSO-based memetic algorithm [5]. The survey of Vansteenwegen et al. [25] gives a review of the most important contributions on the orienteering literature.

Recently, the Orienteering Problem with Time Windows (OPTW) and the Team Orienteering Problem with Time Windows (TOPTW) have been the interest of many researchers. They are considered as the generalization of OP and TOP with the additional time constraints. In these problems, the service of a

© Springer International Publishing Switzerland 2014
P. Legrand et al. (Eds.): EA 2013, LNCS 8752, pp. 183–194, 2014.
DOI: 10.1007/978-3-319-11683-9_15

customer must be started within a time window $[e_i, l_i]$ defined by customer i. The vehicle cannot arrive earlier than time e_i and no later than time l_i. A vehicle arriving earlier than the earliest service time of a customer will incur waiting time. The first who considered the time windows in the OP were Kantor and Rosenwein [8]. They solved the problem with a tree heuristic that was more efficient than the classical insertion heuristics. The only exact method that we found was developed by Righini and Salani [17]. The computational time required by their method to solve large problem instances was very expensive. Therefore, most of the researchers focus on developing approximate methods. Montemanni and Gambardella [12] used ant colony optimization to solve the problem, while Vansteenwegen et al. [26] present an iterated local search metaheuristic. In this method, an insert step is combined with a shake step to explore the search space more efficiently. Tricoire et al. [23] defined the Multi-Period Orienteering Problem with Multiple Time Windows (MuPOPTW) as a new problem for scheduling the customer visits of sales representatives. The MuPOPTW is a generalization of OPTW and TOPTW, where customers may be visited on different days, and may have several time windows for each given day. They propose an exact algorithm embedded in a variable neighborhood search method and provide experimental results for their method on standard benchmark of OPTW and TOPTW instances. Lin and Yu [11] presented a simulated annealing based heuristic approach to solve TOPTW. The method proposed by Labadie et al. [10] combines greedy randomized adaptive search procedure (GRASP) with evolutionary local search (ELS). ELS generates multiple distinct child solutions that are further improved by a local search procedure, while GRASP provides multiple starting solutions to ELS. Labadie et al. [9] introduced granular variant to a VNS algorithm in order to improve its efficiency. Firstly, each arc is evaluated with new cost taken into account traveling times, waiting times and profits. Then, an assignment problem is optimally solved and intervals of granularity are created. These intervals determine subset of promising arcs which will be considered during the node sequences construction in the local search procedure.

In this paper, a metaheuristic-based memetic algorithm (MA) is presented for TOPTW. The proposed MA works with permutation encoding and uses an adapted procedure to optimally split a sequence into a set of routes. The rest of the article is organized as follows. The next section is devoted to the formulation of TOPTW. Section 3 presents the detailed description of the proposed method including the solution representation, the optimal split procedure, and other components and parameters. In Sect. 4, the effectiveness of the proposed algorithm is demonstrated by many computational results based on some benchmark problems. The conclusions are discussed in the final Sect. 5.

2   Formulation of the Problem

TOPTW is modeled with a graph $G = (V, E)$, $V = \{0, 1, 2, ..., n\}$ is the set of vertices where $i \neq 0$ represents a customer and 0 represents the depot. $E = \{(i, j) : i \neq j, i, j \in V\}$ is the edge set. Each vertex $i \in V, i \neq 0$ is associated with a profit P_i and a service time T_i. The visit of a vertex i can start only within a predefined time window $[e_i, l_i]$. The vehicle v cannot arrive later than the time l_i and if it arrives earlier than e_i, it must wait W_i^v before the service can start. Each edge $(i, j) \in E$ is associated with a travel cost $c_{i,j}$ which is assumed to be symmetric and satisfying the triangle inequality. A *tour* R is represented as an ordered list of q customers from V, so $R = (R[1], \ldots, R[q])$. Each tour begins and ends at the depot vertex. We denote the total profit collected from a tour R as $P(R) = \sum_{i=1}^{i=q} P_{R[i]}$, and the total travel cost or duration $C(R) = c_{0,R[1]} + \sum_{i=1}^{i=q-1} c_{R[i],R[i+1]} + \sum_{i=1}^{i=q} W_{R[i]}^R + \sum_{i=1}^{i=q} T_{R[i]} + c_{R[q],0}$. A tour R is feasible if $C(R) \leq l_0$, l_0 being a latest possible arrival time to the depot, and if each customer is serviced within its time window. The fleet is composed of m identical vehicles. A *solution* S is consequently a set of m (or fewer) feasible tours R in which each customer is visited at most once. The goal is to find a solution S such that $\sum_{R \in S} P(R)$ is maximized. For mixed integer linear programming formulations of TOPTW see [12, 26].

3   Memetic Algorithm

Memetic algorithm is a combination of an evolutionary algorithm and local search framework [13]. The basic idea behind memetic approaches is to combine the advantages of the crossover that discovers unexplored promising regions of the search space, and local optimization that finds good solutions by concentrating the search around these regions. The proposed memetic algorithm is based on permutation encoding. The key feature of the proposed method is the split procedure that allows a reduction of the solution space exploration within the global optimization. We introduce an interesting way to represent solutions of TOPTW, known as giant tours. Each giant tour is in fact a neighborhood of solutions in the search space from which the optimal associated solution can be easily extracted by an evaluation process. Therefore, a heuristic using this representation explores a smaller solution space without any loss of information and has a better chance to reach the global optimum. The good results obtained on several extensions of the routing problems have raised a growing attention on the split strategy [16]. Next, all the details of MA implementation are presented.

3.1   Chromosome and Evaluation

The representation of our chromosome consists of an ordered list of all accessible customers in V called a *giant tour*. The giant tour is a permutation of n positive integers, such that each integer corresponds to a customer without trip delimiters. We try to extract m tours from the giant tour while respecting the order of the customers in the sequence. A tour from a permutation π is identified by its starting point i in the sequence and the number of customers following i. A chromosome is evaluated using a tour splitting procedure which optimally partitions π into feasible routes. Using this strategy, the MA searches the set of possible giant tours to find one that gives an optimal solution after splitting.

Bouly et al. [2] proposed an optimal splitting procedure which is specific to TOP. In their method, only tours of maximum length are considered. This means that all customers following i in the sequence are included in the tour as long as all constraints are satisfied, or until the end of the sequence is reached. Such a tour is called *saturated* tour. They proved that solutions containing only saturated tours are dominant. Therefore, only saturated tours were considered in their procedure. Later, Dang et al. [6] introduced a new evaluation procedure in which the limited number of saturated tours is exploited more efficiently to reduce the complexity of the evaluation process. Before reviewing the main idea of the split procedure, we recall the definition of an interval graph [22] as follows. A graph $G = (V, E)$ is an interval graph if there is a mapping I between the vertex set of G and a collection of intervals in the real line such that two vertices of G are adjacent if their respective intervals intersect. Then, for all i and j of V, $[i, j] \in E$ if and only if $I(i) \cap I(j) \neq \emptyset$.

We have extended the split procedure for TOPTW to tackle time windows. When defining a saturated tour R starting with x, we should make sure that each customer is served within its time window and that $C(R) \leq l_0$, where l_0 is the latest possible arrival time to the depot and $C(R)$ the total travel duration. So, a waiting time is added each time the vehicle arrives at a customer before the beginning of his time window. The set of extracted tours from a giant tour can be mapped to the set of vertices of an interval graph X. An edge in X indicates the presence of shared customers between the associated tours. A split procedure looks for m tours without any shared customer such that the sum of their profit is maximized. So this is equivalent to solve a knapsack problem with the conflict graph X. In this particular knapsack problem, the number of items is equal to the number of possible tours. This number is equal to n when only saturated tours are considered. The weight of each item is one and the capacity of the knapsack is m. Such a problem can be solved in $O(m \cdot n)$ time and space [18].

Proposition 1. *Given a TOPTW instance where m is the maximum number of available vehicles and π a permutation of n customers, the split procedure of π can be done optimally in $O(m \cdot n)$ time and space.*

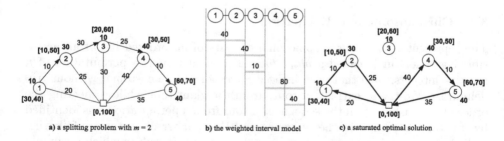

a) a splitting problem with $m = 2$ b) the weighted interval model   c) a saturated optimal solution

Fig. 1. An example of splitting problem.

The evaluation process is performed with dynamic programming technique: A two-dimensional array of size $m \cdot n$ is used to memorize the maximum reachable profit during process. Then, a backtrack is performed in order to determine the tours corresponding to the optimal solution. The first graph of Fig. 1 shows a sequence $S = (1, 2, 3, 4, 5)$ where each customer has a profit and a time window given in the square brackets. To simplify, we assume that the service times are set to 0, the number of the vehicles m used is equal to 2, and the maximum operation time l_0 is 100. The interval model is given in the Fig. 1.b. The first interval [1,2] for example with weight 40 corresponds to the collected profit of the trip $(d, 1, 2, a)$. Vehicle leaves the depot at time 0, waits 10 units of time at node 1 before leaving it to serve node 2 at time 40. The customer 3 cannot be included in the trip, since its time window is already closed when the vehicle reaches it at time 70. The other intervals $[i, j]$ of the graph are similarly defined. The maximum score obtained in the solving steps is equal to 120. Finally, we give the optimal solution obtained by the algorithm in Fig. 1.c. It is composed of two tours starting respectively with customers 1 and 4.

3.2 Population

A small part of the initial population is created with a fast heuristic procedure and the remainder is generated randomly. In the proposed Iterative Destruction/Construction Heuristic (IDCH), we build a feasible solution by inserting at every iteration an unrouted customer. This process is performed using Best Insertion Algorithm (BIA). Initially, IDCH removes a limited random number of customers $D \in \{1, 2, 3\}$ from the current solution. Then, the travel cost of tours is reduced using 2-opt* and Or-opt exchanges [14]. In the next step, we rebuild the solution by re-inserting unrouted customers in all possible ways. To ensure that the feasibility of an insertion is verified in $O(1)$, we record for each already included customer i in a route r, its waiting time W_i^r and the maximum delay allowed for the service $Maxshift_i^r$. All feasible insertions of each unserved customer u between two couple of adjacent customers i and j are evaluated. This is done according to a suitable cost function $f(u) = Shift_u/(P_u)^\alpha$ where $Shift_u = (c_{i,u} + W_i^r + T_u + c_{u,j} - c_{i,j})$. The feasible insertion that minimizes the cost is then processed. In addition, priority coefficient $prio_u$ is associated to each customer u. Whenever the customer is not routed through a construction phase its priority is increased by the value of its profit. The customer u that has a lager $prio_u$ is more likely to be inserted. When a limited number of iterations $iter_{perturb} = n$ is reached without a strict improvement, a method of diversification is performed. Diversification stands for random moves that can deteriorate the current solution by removing a large number of customers $D_{perturb} \in [1, n/m]$. The destruction and construction phases are iterated until $iter_{max} = n^2$ iterations without improvement.

3.3 Selection and Crossover

In this work, the binary tournament method [15] is adopted to select a couple of parents among the population. Two chromosomes are randomly selected in the population, and the chromosome with the best evaluation becomes the first parent. The tournament is repeated for the second parent. These parents are then combined using linear order crossover or LOX [15]. LOX first chooses two cut points randomly and passes the section enclosed by the cut points from one parent to child. Then, the unpassed customers are placed in the unfilled positions using the order of their occurrence in the other parent.

3.4 Local Search Engine

When a new child is computed with the crossover operator, the local search scheme is applied with a probability pm. Neighborhoods are selected in a random order. The search in a given neighborhood is stopped as soon as a better solution is found. Then, a new neighborhood is chosen randomly. This process is stopped when all neighborhoods fail to bring out an improvement to the current solution. The set of local search operators used in the Memetic Algorithm are:

- *2-opt* operator*: two routes r_1 and r_2 are divided into two parts. Then the first part of r_1 is connected to the second part of r_2, while the first part of r_2 is connected to the second part of r_1.
- *Or-opt operator*: consider a sequence of one, two or three consecutive customers in the current solution, and move the sequence to another location in the same route.
- *destruction/repair operator*: first, a random number of customers (between 1 and $\frac{n}{m}$) is removed from an identified solution. Then, the lowest possible insertion cost $\frac{Shift_i}{(P_i)^\alpha}$ of each unrouted customer i is evaluated. The visit with the lowest ratio will be selected for insertion.
- *shift operator*: a customer is removed from its current position and is relocated at another one. Every possible insertion position for every customer is considered.
- *swap operator*: positions of every two customers in the sequence are exchanged.

3.5 Population Update

When an offspring solution s_{new} is created by the crossover operator presented in Sect. 3.3 and improved by the local search algorithm described in Sect. 3.4, we decide if the improved offspring should be inserted into the population and which existing solution of the population should be replaced. Basically, our decision is made based on both: the solution quality and the distance between solutions in the population. The update procedure is applied if the performance of new solution s_{new} is better than the worst individual. Population is a list of solutions sorted in descending order according to two criteria: the total collected profit and the travel cost/time. Two solutions are said to be similar or identical if the evaluation procedure returns the same profit and a difference in travel cost/time lower than a value δ. If there is a solution s similar to s_{new}, then s is replaced with s_{new}. Otherwise the worst individual is deleted and the new solution is inserted into the population.

3.6   Basic Algorithm

The proposed Memetic Algorithm associates all the elements described above. Algorithm 1 presents a synthetic view of the whole process. The algorithm starts with an initial set of solutions, called population. During each iteration, two parents are selected and crossover operator is applied to create a new solution. The obtained child chromosome has a probability pm of being mutated using a set of local search techniques repeatedly. Finally, it is inserted within the population according to its fitness evaluation. The stopping criterion for MA is after reaching a maximum number of iterations without improvement. That is to say after reaching the number of iterations where the child chromosome has the same fitness as an existing chromosome in the population, or when its evaluation is worse than the worst chromosome in the current population.

Algorithm 1. Basic algorithm

    Data: POP a population of N solutions;
    Result: $S_{POP}[N-1]$ best solution found;
1 **begin**
2   initialize and evaluate each solution in POP (see Sect. 3.2);
3   **while** $NOT(stopping\ condition)$ **do**
4   select 2 parents $POP[p1]$ and $POP[p2]$ using binary tournament;
5         $C \leftarrow LOX(POP[p1], POP[p2])$;
6   **if** $rand(0,1) < pm$ **then**
7   apply local search on C (see Sect. 3.4);
8   **if** $f(C) \geq f(POP[0])$(see Sect. 3.5) **then**
9   **if** $\nexists p \| (f(POP[p]) = f(C))$ **then**
10   eject $POP[0]$ from POP ;
11   reset stopping condition ;
12   **else**
13   update stopping condition;
14   insert or replace C in right place in POP ;
15   **else**
16   update stopping condition;

---

4   Numerical Results

We used 56 instances designed by Solomon [19] and 20 instances designed by Cordeau et al. [4] to test our new proposed algorithm. Solomon's 100-customer instances are divided into *random, clustered* and *randomclustered* categories. In Cordeau's instances, the number of customers varies between 48 and 288. A third set of benchmark was introduced by Vansteenwegen et al. [26] using the original instances of Solomon and Cordeau. In these instances, the number of vehicles considered allows to visit all customers that is why the optimal solutions of these instances are known since they are equal to the sum of customers' profits. Travel time between two customers is assumed to be equal to the travel distance. It is rounded down to the first decimal for the Solomon's instances and to the second decimal for the Cordeau's instances. The whole algorithmic approach was implemented in C++ using the Standard Template Library (STL) for data structures and was compiled using the GNU GCC compiler on an AMD Opteron 2.60 GHz in a Linux environment.

4.1   Parameter Setting

A number of different alternative values were tested and the ones selected are those that gave the best computational results concerning both the quality of the solution and the computational time needed to achieve this solution. When the population is initialized, 5 chromosomes are generated by the IDCH heuristic and the rest (35) are generated randomly. The similarity measurement of individuals δ is set to 0.01 and the local search rate pm is calculated as: $1 - \frac{iter}{itermax}$ where $iter$ is the number of consecutive iterations without improvement. The algorithm stops when $iter$ reaches $itermax = k * n/m$. The cost function $C(u) = Shift_u/(P_u)^\alpha$ of the BIA heuristic uses a random value of α generated in $[1, 3]$. This control parameter makes our IDCH less predictable and actually a randomized heuristic. Moreover, the score becomes more relevant than the time consumption when deciding which unrouted client is the most promising to insert. If α is set to 1, the obtained results are worse. Finally only two parameters are required to be tuned, they are the stopping condition k and the population size N. Computational experiments were conducted on a representative subset of the problem characteristics (problem size, distribution of customer location, and time windows characteristic). This small subset includes 40 instances: 6 problems from Solomon's instances and 4 problems from Cordeau's instances with $m = 1, 2, 3, 4$. The value of k and N were varied from 10 up to 50 with steps of 10. This results 25 different (k, N) settings to be tested. The algorithm was run 5 times on different randomly generated seeds for each instance. For an overall performance comparison between different configurations, we use two following measures. The first one is the relative gap to the best known solutions, denoted $rpe(\%)$ and the second is the average computational time in seconds CPU_{avg}. The results for each of the 25 parameter combinations tested are illustrated in Fig. 2. We adopt the parameter settings $(10, 40)$ which gives a good trade off between algorithm performance and computational time.

Table 1. Performance comparison based on RPE average for each data set of the standard benchmark.

Instance set	ACS		ILS		VNS		GRASP-ELS		SA		GVNS		MA	
	rpe	cpu_{avg}	rpe	cpu_{avg}	rpe	cpu_{avg}	rpe	cpu_{avg}	rpe	cpu_{avg}	rpe	cpu_{avg}	rpe	cpu_{avg}
m=1														
c100	0	6.34	1.11	0.33	0	98.39	0	22.59	0	21.07	0.56	166.46	0	0.98
r100	0	383.40	1.90	0.19	0	89.10	0.11	3.51	0.11	23.34	1.72	29.43	0	5.38
rc100	0	143.21	2.92	0.23	0	65.21	0.33	1.99	0	22.19	1.88	9.80	0	1.59
c200	0.40	342.61	2.28	1.71	0	560.17	0.40	32.18	0.13	37.49	0.55	192.40	0	122.40
r200	2.18	1556.70	2.89	1.66	0.40	1065.82	0.59	11.18	1.29	45.83	2.44	33.82	**-0.52**	236.10
rc200	1.23	1544.55	3.43	1.63	0.07	869.41	1.37	8.21	0.96	50.25	2.53	16.01	**-0.02**	201.52
pr01-pr10	1.05	1626.61	4.72	1.75	0	822.07	0.73	5.03	0.97	112.21	0.54	12.37	**-0.02**	485.98
pr11-pr20	10.73	887.66	9.11	1.98	0.93	1045.93	1.70	7.90	3.25	162.40	2.71	24.22	**0.39**	903.08
m=2														
c100	0.15	818.00	0.94	1.08	0	87.98	0	70.94	0	26.42	0.47	139.53	0	70.09
r100	0.34	1559.36	2.27	0.87	0.06	63.46	0.92	7.97	0.14	36.63	1.10	60.34	**-0.12**	45.98
rc100	0.38	1375.78	2.47	0.71	0.23	55.16	1.46	4.66	0.19	40.48	0.78	20.31	0	46.33
c200	1.27	1398.10	2.54	3.46	0.51	545.65	0.09	29.26	1.18	53.66	0.25	33.79	0	164.93
r200	3.11	2735.15	2.69	2.27	0.20	1015.08	0.28	17.58	0.53	91.40	0.62	14.73	**-0.57**	634.67
rc200	2.64	2342.72	4.08	2.20	0.43	804.83	0.59	17.14	1.18	80.10	1.62	12.76	**-0.60**	355.97
pr01-pr10	2.35	1889.66	5.99	4.76	0.63	524.83	0.87	19.46	2.21	173.93	0.57	39.09	**-0.44**	1291.54
pr11-pr20	4.79	2384.81	7.65	5.21	1.04	618.78	2.21	28.77	3.66	201.63	0.98	82.44	**-0.24**	2144.27
m=3														
c100	0.11	1043.24	2.44	1.50	0	85.49	0.13	86.74	0.22	35.26	0.34	165.01	0	70.77
r100	0.55	1668.86	1.78	1.67	0.21	61.91	0.89	13.86	0.38	56.07	1.21	73.93	**-0.01**	58.56
rc100	1.19	1476.81	3.14	1.11	0.36	60.62	1.83	8.65	0.64	42.80	0.91	33.68	**-0.01**	54.72
c200	0.55	1413.11	1.98	2.08	0.16	196.80	0.45	26.75	0.35	53.93	0.64	55.42	**-0.10**	104.73
r200	0.13	1171.65	0.30	1.36	0.03	321.65	0	2.49	0.08	41.95	0.11	6.97	0	**74.22**
rc200	0.37	1607.85	1.37	1.73	0.04	404.01	0.06	8.34	0.20	58.98	0.25	7.41	**-0.07**	212.43
pr01-pr10	3.01	2163.80	6.57	9.24	1.50	473.20	1.31	40.55	2.33	197.01	0.35	85.90	**-0.33**	1416.21
pr11-pr20	5.19	2383.29	8.91	9.69	1.48	517.48	2.00	42.95	3.51	251.83	0.72	150.73	**-0.71**	2388.19
m=4														
c100	0.47	1056.05	2.93	2.57	0.09	81.87	0.50	84.58	0.36	49.51	0.85	133.22	**-0.19**	106.15
r100	0.99	1652.54	3.25	2.60	0.24	61.17	0.88	24.18	0.67	58.38	1.15	84.74	**-0.11**	79.46
rc100	0.92	1854.00	3.07	1.98	0.34	58.47	1.43	13.35	0.26	68.13	0.85	36.91	**-0.24**	57.66
c200	0	7.70	0	1.00	0	104.78	0	0.01	0	41.76	0	0.55	0	0.04
r200	0	126.46	0	0.87	0	150.74	0	0.03	0	39.71	0	0.27	0	0.10
rc200	0	646.72	0	1.24	0	164.56	0	0.03	0	40.15	0	0.88	0	0.15
pr01-pr10	2.34	2447.70	6.63	14.07	1.40	403.17	1.42	45.75	1.76	255.57	0.60	127.33	**-1.12**	1807.40
pr11-pr20	4.18	2583.50	7.16	13.74	0.90	408.01	1.20	65.33	2.57	283.98	0.64	232.64	**-2.23**	2784.70
Average	1.65	1401.79	3.38	3.09	0.36	375.62	0.74	22.60	0.96	88.30	0.87	64.34	**-0.23**	524.00

Table 2. Performance comparison based on ARPE average for each data set of the new benchmark.

Instance set	ILS		GRASP-ELS		SA		GVNS		MA	
	$arpe$	cpu_{avg}	$arpe$	cpu_{avg}	$arpe$	cpu_{avg}	$arpe$	cpu_{avg}	$arpe$	cpu_{avg}
new Solomon's instances	1.12	2.38	0.35	70.34	0.30	35.70	0.65	16.92	**0.04**	43.02
new Cordeau's instances	2.32	30.41	1.04	565.98	0.92	71.48	1.25	51.34	**0.76**	112.63
Average	1.72	16.40	0.70	318.16	0.61	53.59	0.95	34.13	**0.40**	77.82

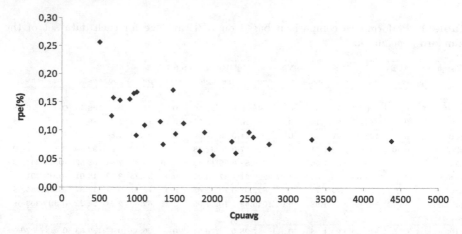

Fig. 2. Pareto front solutions obtained with different settings of the stopping condition k and the population size N

4.2 Performance Comparison

In order to investigate the performance of the proposed MA for TOPTW, we compare it with: the Ant Colony System (ACS) of [12], the Iterated Local Search (ILS) of [26], the Variable Neighborhood Search (VNS) of [23], the Simulated Annealing approach (SA) of [11], the Greedy Randomized Adaptive Search procedure of (GRASP-ELS) [10] and the Granular Variable Neighborhood Search (GVNS) of [9]. The results of GVNS, GRASP-ELS and ACS were obtained with 5 runs of the algorithm on each instance. VNS was run 10 times per instance while ILS and SA were executed only once. We used the same protocol as in the state-of-the-art methods and run MA 5 times for each instance. The quality of the produced solutions is given in terms of the relative percentage error (RPE) for the standard benchmark and in terms of the average relative percentage error (ARPE) for the new data set where there exists a solution visiting all customers. Tables 1 and 2 summarize the comparison and report the percentage error (RPE or $ARPE$) and the average computational time in seconds CPU_{avg} for each instance set. MA produces the best relative gap which is equal to $-0,23\%$ for the standard benchmark and $0,40\%$ for the new data set. The first conclusion that can be drawn from these tables is that MA is very competitive compared to the others methods. It outperforms the other methods and improves 101 instances for which the optimal solution remains unknown. However, one should note that MA is far more time consuming. Actually, on the largest instances. MA needs more time to get good quality solutions. The reason appears to be that a lot of time is consumed by local-search operators. This is necessary to take entirely advantage of the MA component.

5 Conclusion

In this paper, a Memetic Algorithm was proposed for the Team Orienteering Problem with Time Windows. The key feature of our algorithm is the use of an Optimal Split procedure especially intended for TOPTW that runs in $O(m \cdot n)$. The proposed algorithm integrates several optimization methods, including heuristic approaches, a crossover operator, a local search optimization procedure and a quality-and-diversity based population updating strategy. The computational results obtained prove the efficiency of our memetic algorithm for TOPTW in comparison with the existing ones. The algorithm brings further improvements and has allowed the identification of new best known solutions. The method is also very flexible in the sense that it can address many problem variants with a unified methodology and common parameter settings. Future work will focus on extending the methodology to a wider array of vehicle routing problems with time windows.

References

1. Archetti, C., Hertz, A., Speranza, M.G.: Metaheuristics for the team orienteering problem. J. Heuristics **13**(1), 49–76 (2007)
2. Bouly, H., Dang, D.C., Moukrim, A.: A memetic algorithm for the team orienteering problem. 4OR **8**(1), 49–70 (2010)
3. Chao, I.M., Golden, B., Wasil, E.A.: The team orienteering problem. Eur. J. Oper. Res. **88**, 464–474 (1996)
4. Cordeau, J.F., Gendreau, M., Laporte, G.: A tabu search heuristic for periodic and multi-depot vehicle routing problems. Networks **30**(2), 105–119 (1997)
5. Dang, D. C., Guibadj, R.N., Moukrim, A.: A PSO-based memetic algorithm for the team orienteering problem. In: Di Chio, C., et al. (eds.) EvoApplications 2011, Part II. LNCS, vol. 6625, pp. 471–480. Springer, Heidelberg (2011)
6. Dang, D.C., Guibadj, R.N., Moukrim, A.: An effective pso-inspired algorithm for the team orienteering problem. Eur. J. Oper. Res. **229**(2), 332–344 (2013)
7. Golden, B., Levy, L., Vohra, R.: The orienteering problem. Nav. Res. Logistics **34**, 307–318 (1987)
8. Kantor, M.G., Rosenwein, M.B.: The orienteering problem with time windows. J. Oper. Res. Soc. **43**(6), 629–635 (1992)
9. Labadie, N., Mansini, R., Melechovský, J., Calvo, R.W.: The team orienteering problem with time windows: An lp-based granular variable neighborhood search. Eur. J. Oper. Res. **220**(1), 15–27 (2012)
10. Labadie, N., Melechovský, J., Calvo, R.W.: Hybridized evolutionary local search algorithm for the team orienteering problem with time windows. J. Heuristics **17**(6), 729–753 (2011)
11. Lin, S.W., Yu, V.F.: A simulated annealing heuristic for the team orienteering problem with time windows. Eur. J. Oper. Res. **217**(1), 94–107 (2012)
12. Montemanni, R., Gambardella, L.: Ant colony system for team orienteering problems with time windows. Found. Comput. Decis. Sci. **34**(4), 287–306 (2009)
13. Moscato, P.: Memetic algorithms: a short introduction. In: New Ideas in Optimization, pp. 219–234. McGraw-Hill Ltd., Maidenhead (1999)
14. Potvin, J.Y., Kervahut, T., Garcia, B.L., Rousseau, J.M.: The vehicle routing problem with time windows part i: Tabu search. INFORMS J. Comput. **8**(2), 158–164 (1996)
15. Prins, C.: A simple and effective evolutionary algorithm for the vehicle routing problem. Comput. Oper. Res. **31**(12), 1985–2002 (2004)

16. Prins, C., Labadie, N., Reghioui, M.: Tour splitting algorithms for vehicle routing problems. Int. J. Prod. Res. **47**(2), 507–535 (2009)
17. Righini, G., Salani, M.: Decremental state space relaxation strategies and initialization heuristics for solving the orienteering problem with time windows with dynamic programming. Comput. Oper. Res. **36**(4), 1191–1203 (2009)
18. Sadykov, R., Vanderbeck, F.: Bin packing with conflicts: a generic branch-and-price algorithm (2012). (preprint accepted for publication in INFORMS Journal on Computing)
19. Solomon, M.M.: Algorithms for the vehicle routing and scheduling problems with time window constraints. Oper. Res. **35**(2), 254–265 (1987)
20. Souffriau, W., Vansteenwegen, P., Vanden Berghe, G., Van Oudheusden, D.: A path relinking approach for the team orienteering problem. Comput. Oper. Res. **37**(11), 1853–1859 (2010)
21. Tang, H., Miller-Hooks, E.: A tabu search heuristic for the team orienteering problem. Comput. Oper. Res. **32**, 1379–1407 (2005)
22. Tarjan, R.E.: Graph theory and gaussian elimination. Technical report, Stanford University (1975)
23. Tricoire, F., Romauch, M., Doerner, K.F., Hartl, R.F.: Heuristics for the multi-period orienteering problem with multiple time windows. Comput. Oper. Res. **37**, 351–367 (2010)
24. Tsiligirides, T.: Heuristic methods applied to orienteering. J. Oper. Res. Soc. **35**(9), 797–809 (1984)
25. Vansteenwegen, P., Souffriau, W., Van Oudheusden, D.: The orienteering problem: a survey. Eur. J. Oper. Res. **209**(1), 1–10 (2011)
26. Vansteenwegen, P., Souffriau, W., Vanden Berghe, G., Van Oudheusden, D.: Iterated local search for the team orienteering problem with time windows. Comput. Oper. Res. **36**(12), 3281–3290 (2009)

Genetic Programming

Learning Selection Strategies
for Evolutionary Algorithms

Nuno Lourenço[1]([✉]), Francisco Pereira[1,2], and Ernesto Costa[1]

[1] Department of Informatics Engineering, CISUC, University of Coimbra,
Polo II - Pinhal de Marrocos, 3030 Coimbra, Portugal
{naml,ernesto,xico}@dei.uc.pt
[2] Instituto Politcnico de Coimbra, ISEC, DEIS, Rua Pedro Nunes,
Quinta da Nora, 3030-199 Coimbra, Portugal

Abstract. Hyper-Heuristics is a recent area of research concerned with the automatic design of algorithms. In this paper we propose a grammar-based hyper-heuristic to automate the design of an Evolutionary Algorithm component, namely the parent selection mechanism. More precisely, we present a grammar that defines the number of individuals that should be selected, and how they should be chosen in order to adjust the selective pressure. Knapsack Problems are used to assess the capacity to evolve selection strategies. The results obtained show that the proposed approach is able to evolve general selection methods that are competitive with the ones usually described in the literature.

1 Introduction

Evolutionary Algorithms (EAs) are computational methods loosely inspired by the principles of natural selection and genetics, that have been successfully applied over time to complex problems involving optimization, learning or design. EAs work by defining an initial population of candidate solutions to the problem, which are then iteratively improved by means of variation operators. The subset of individuals that undergo the modification process must be selected according to some fitness criteria. The quality of the solutions achieved by the EA depend on the careful adjustment of some its components and/or parameters. The design is usually performed off-line, by hand, and requires the use of expertise knowledge.

Hyper-Heuristics (HH) is a recent area of research, involving the construction of specific, high-level, heuristic problem solvers, by searching the space of possible low-level heuristics for the particular problem one wants to solve [10]. HH can be divided in two major groups [1]: the selection group encompasses HH that search for the best sequence of low-level heuristics, selected from a set of predefined methods usually applied to the problem one intends to solve; the other group includes methods that promote the creation of new heuristics. In the later case, the HH iteratively learns the specific algorithm which is then applied to solve the

P. Legrand et al. (Eds.): EA 2013, LNCS 8752, pp. 197–208, 2014.
DOI: 10.1007/978-3-319-11683-9_16

problem at hand. During this process, the HH are usually guided by feedback obtained through the execution of each candidate solution in instances of the problem that needs to be solved. Genetic Programming (GP), a branch of EAs, has been increasingly adopted as a HH to search for effective problem solving algorithmic strategies [3, 8]. In the recent years, Grammatical Evolution (GE) [7], a form of GP, has been used with success as a HH, since it allows the enforcement of semantic and syntactic restrictions, by means of a grammar.

In this paper we propose and test a GE-based HH framework to evolve a EA particular component. Specifically, we propose a framework to evolve the selection mechanism used by the EA. With this goal in mind, we expect to obtain selection mechanisms that are general, and are able to successfully guide EAs to solve the problem at hand. The selection component is important for the success of the algorithm, since it determines which individuals should be combined to produce new solutions. We describe a set of experiments, where we show that the framework is able to evolve selection algorithms that are competitive with the ones commonly used in EAs. Moreover we investigate the generalization capacity of the evolved algorithms, by applying them to unseen scenarios. The results are statistical validated.

The paper is organized as follows. In Sect. 2 we discuss some previous relevant work on HH for nature-inspired algorithms, and present the grammar used to evolve selection strategies. In Sect. 3 we introduce the experimental setup for the learning phase and present the results. Section 4 deals with the validation and generalization of the learned selection strategies. In Sect. 5 we summarize the results and suggest directions for future work.

2   A Grammatical Evolution Hyper-Heuristic

The proposed HH relies on GE to search for selection methods. GE is a GP branch, more specifically a form of Grammar-based GP, in which the variation operators are applied to solutions encoded as binary strings. A mapping process is then required to decode this information into an executable algorithmic strategy. The mapping is done by means of a grammar and this process decouples the search engine from the evaluation mechanism. For these reasons, a GE system is general and flexible [7].

2.1   Grammar Definition

To apply a GE engine in our HH framework we must define a grammar whose words are specific selection strategies. In this work we propose a grammar with some modifications to the traditional Backus-Naur Form (BNF), inspired by [2]. These modifications aim to overcome some limitations that the BNF imposes, namely the lack of tools to allow repetition of non-terminal symbols and ranges of alternative values. The first extension is the addition of the operator $\sim$ to signal the repetition of non-terminals. The full syntax is as follows: $\sim< a >< NT >$, where $< a >$ is an integer or terminal value, indicating that the non-terminal

$< NT >$ should be repeated $< a >$ times. The second extension is the addition of valued range alternatives. A range of numeric alternative values can be compactly specified, using the operator &. Thus $< int >::= 0\&5$ is equivalent to $< int >::= 0|1|2|3|4|5$. Taking these extensions into account, the grammar used in this work is as follows:

```
<start> ::= <calculate-parents> <selection-strategy>

<selection-strategy> ::= parents = {~number-of-parents<elements>}

<elements> ::= get_rank(<rank>)

<rank> ::= 0 & POP_SIZE

<calculate-parents> ::= number-of-parents = (random01() * POP_SIZE)
```

The $< start >$ symbol represents the grammar axiom. The grammar starts by calculating the number of parents that the strategy should select, according to a percentage of the total individuals available (POP_SIZE). The evolved strategies are targeted for EAs with a crossover operators, thus we enforce an even number of parents in the selection pool. Afterwards, a selection strategy to choose which individuals will appear in the selection pool is generated. The solutions from the current population are ranked by fitness and a selection strategy emerges by defining which ranks should be chosen as parents.

2.2 Related Work

Several efforts have been reported in literature to automatically evolve nature-inspired algorithms. In [11], Tavares et al. adopted GP to evolve a population of mapping functions between the genotype and the phenotype. Experimental results showed that GP finds mapping functions that can obtain results as good as the ones that are designed by hand.

In [3], Keller et al. propose a linear-GP HH to evolve heuristics to Travelling Salesman problem. In their work they propose several small languages to reduce the search space size. They conclude that the proposed HH is able to evolve heuristics that are able to solve the problem at hand, and that they are parsimonious, i.e. the heuristics make a good use of the resources available.

In [12], Tavares et al. proposed a GE framework to evolve Ant Colony Optimization Algorithms (ACO) to the Traveling Salesman Problem. The results showed that the proposed framework is able to evolve ACO algorithms that are competitive with the human designed ACOs.

Lourenço et al. [5] proposed a GE based HH to evolve full-featured EAs. The results showed that the proposed architecture is able to evolve effective algorithms for the problems under consideration.

In [13] Woodward et al. propose an HH to evolve mapping rules that assign fitness values to each individual in the population. These fitness values are then used to select individuals, using a fitness-proportional mechanism. They consider a set of transformations that can be applied to either the rank or the fitness, and then return the new fitness value of each individual. The experiments results conducted showed that the evolved strategies are human competitive.

3 Learning Selection Heuristics

In this section we aim to gain insight into the capacity of the proposed HH to evolve effective selection strategies. The settings adopted by the GE-based HH for all the tests conducted are depicted in Table 1. Individuals evolved by the HH encode potential parent selection strategies. To estimate their relevance, one must access how they help an EA to solve a given problem. Therefore, each HH individual is implanted in a standard EA, which in turn will solve an instance of an optimization problem. The quality of the best solution found by this EA is used to assign fitness to the corresponding evolved selection strategy. Running an EA to assign fitness to each evolved selection strategy is a computational expensive task. To minimize the computational overhead, we rely on the following conditions to assess the quality of the evolved strategies: (i) one single instance of moderate size is used to assign fitness; (ii) only one run is performed.

We report experiments using three different EA settings as surrogates for the selection strategies. In all of them, the maximum population size (POP_SIZE) is set to 50 and the number of generations is set to 250. Three possible replacement strategies, $R1$, $R2$, and $R3$, are considered (see Table 2). $R1$ corresponds to a standard generational EA, whereas the last two implement a steady-state architecture where descendants compete with existing individuals for survival based on the fitness criterion. Both $R1$ and $R2$ force the evolved selection strategies to select a number of parents that is equal to POP_SIZE, thus the grammar production $< calculate - parents >$ simply becomes $< calculate - parents >$ $::=number - of - parents = POP\_SIZE$. On the contrary, $R3$ allows the selection strategy to choose a number of parents that is lower than POP_SIZE. All three replacement strategies consider uniform crossover with a rate of 0.9 and swap mutation with rate 0.01 as variation operators. Additional combinations

Table 1. Parameter setting for the GE-based Hyper-Heuristic

Parameter	Value
One point crossover probability	0.9
Bit flip mutation	0.01
Codon duplication probability	0.01
Codon pruning probability	0.01
Population size	100
Selection	Tournament with size equal 3
Replacement	Steady state
Codon size	8
Number of wraps	3
Codons in the initial population	50–55
Generations	50
Runs	30

Table 2. Replacement strategies used in the surrogate EAs.

Setting	Fixed	Replacement strategy
R1	Yes	Generational
R2		Steady state
R3	No	Steady state

of variations operators were tested with similar outcomes to those reported in this paper (detailed results are not shown due to space constraints).

3.1 The 0-1 Knapsack Problem

The combinatorial optimization *0-1 Knapsack Problem* (KP) was selected as the testbed for our experiments. It can be described as follows: given a set of n items, each of which with some profit p and some weight w, how should a subset of items be selected to maximize the profit while keeping the sum of the weights bounded to a maximum capacity C? In all instances adopted in our study, the knapsack capacity was set to half of the sum of the weights of all items. A standard binary representation is adopted and evaluation considers a linear penalty function to punish invalid solutions [6].

3.2 Results

The KP instance used to evaluate the selection strategies is composed by $n = 100$ items. Table 3 summarizes the results of the off-line learning process. Note that the results are displayed in terms of the normalized root mean squared error. Every cell contains two values: the number of GE runs that discovered selection strategies that helped the EA to discover the optimum (*BestHits*) and the Mean Best Fitness (*MBF*) together with the corresponding standard deviation. The outcomes reveal that, for all training situations, the HH is able to learn effective selection strategies.

A detailed inspection reveals that the replacement strategies used in the surrogate EA lead to the appearance of selection methods with different selective pressure. The three lines from Fig. 1 (one from each replacement strategy) help to clarify this issue. For every setting we selected the best selection algorithm evolved in each run and created charts displaying the distribution of the appearance of the possible ranks (values displayed are averages of 30 runs). Note that rank 0 corresponds to the best individual and rank 49 to the worst. An inspection of the figure shows that selection strategies evolved inside a generational surrogate (R1) have a higher selective pressure than those that evolved in the steady state surrogates. In generational EAs, the whole population is replaced at each generation. The HH acknowledges the risk of losing good quality solutions and promotes the appearance of selection strategies with a high selective pressure, thereby maximizing the likelihood of passing information contained in

Table 3. Hyper-Heuristic learning results (for 30 runs)

	Replacement strategies		
	R1	**R2**	**R3**
Best Hits	30	30	30
MBF	0.000 (±0.000)	0.000 (±0.000)	0.000 (±0.000)

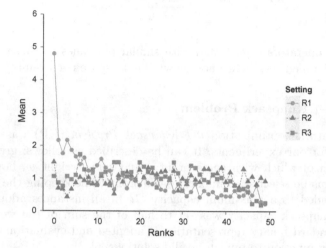

Fig. 1. Rank distribution in the best evolved strategies with the three replacement settings.

good quality solutions to the next generations. On the other hand, in steady state surrogate EAs, the ranks are distributed more or less evenly. This results is not unexpected, since in this scenario, the greedy replacement mechanism already ensures selective pressure: an offspring only enters the population if it is better than its parents. Therefore the selection strategy in these EAs can act more like a diversity preservation mechanism. Finally, in Fig. 2 we exemplify the rank distribution of one of the best evolved strategies, using the $R1$ setting.

4   Validation of the Learned Selection Strategies

The experiments described in this section aim to study how the best strategies discovered by the GE-based HH behave in KP instances that are different from the one used in learning. We selected 4 evolved strategies from each possible replacement strategy, taking into account the following criteria: (i) quality of the solution; (ii) time taken to reach a solution. In the remainder of this section these selection strategies are identified as $R11$ through $R14$ for methods evolved with the $R1$ replacement strategy, $R21$ through $R24$ for $R2$ replacement strategy, and $R31$ through $R34$ for $R3$ replacement strategy.

This experimental study will help to gain insight into the optimization performance of EAs that have the learned strategies as selection methods. Also, we

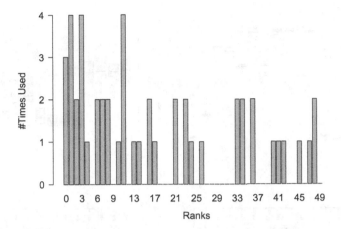

Fig. 2. Example of the rank distribution of a selection strategy evolved with the $R1$ setting.

will verify if the strategies generalize well to unseen instances and are competitive with standard hand-designed selection strategies. Three common selection options (Roulette Wheel, Tournament with size 2, and Tournament with size 3) are considered. We report results obtained with a generational and a steady-state surrogate EAs, both of them relying on uniform crossover with rate 0.9 and binary swap mutation with rate $1/n$ as variation operators. A KP instance with 1000 items and with the knapsack capacity set to half of the sum of the weights of all items was selected for the validation analysis. In every optimization scenario, 30 runs were performed and the best solution found during the execution was recorded. To support our analysis we apply the Friedman's ANOVA test to check for statistical differences in the means. When differences are detected, the *post-hoc* Wilcoxon Signed Rank Test, with Bonferroni correction, is applied to perform the pairwise comparisons. In both tests we used a significance level $\alpha = 0.05$.

Figure 3 presents the MBF box plot distribution of the 15 selected strategies (12 evolved and 3 hand-designed) for each validation scenario: Panel (a) displays the results for the generational surrogate, whereas panel (b) presents the results for the steady-state surrogate. Clearly, the performance of the evolved strategies is related to the configuration where they are applied. Strategies $R11 - R14$ were evolved with a generational EA surrogate and, as a consequence, they promote a considerable selection pressure. Therefore it is not a surprise that these strategies achieve good results in a validation scenario where a generational surrogate is adopted (see Panel 3a). On the contrary, strategies $R21 - R24$ and $R31 - R34$ have a low selective pressure and are inadequate for a generational EA environment.

An opposite situation arises in the steady-state validation surrogate (panel 3b). In this scenario, and given the fitness-based replacement strategy adopted, selection methods evolved in a generational environment converge prematurely

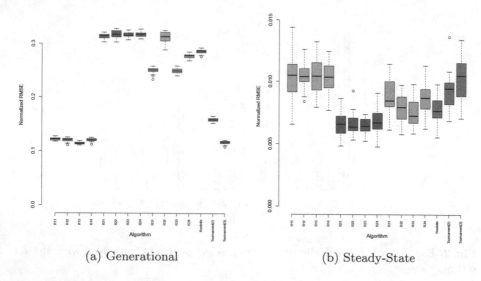

(a) Generational   (b) Steady-State

Fig. 3. Optimization results of the 15 selection strategies chosen for the validation study: panels (a), (b), present the results obtained with the generational and steady state EAs, respectively.

to sub-optimal regions of the search space. The remaining 8 evolved strategies were obtained in a scenario similar to the one used in this validation phase. For that reason they contain features that help to maintain diversity and to effectively help the EA to discover the regions of the search space containing the best solutions. The distinction between these two sets of evolved selection methods confirms that the HH framework is able to generate strategies that are suited to the specific features of the training environment.

The results displayed in the two panels of Fig. 3 confirm that the HH is able to evolve selection methods competitive with the hand-designed approaches. The information displayed in Table 4 helps to further clarify the relative performance of learned strategies. Considering the MBFs attained, we performed a full set of pairwise comparisons between evolved strategies and the hand designed algorithms and present a graphical overview: A $+++$ indicates that the algorithm in the row is statistically better than the one in the column, and that the effect size is large ($r \geq 0.5$). As an example, $R11$ clearly outperforms Roulette Wheel in the Generational surrogate. A $++$ sign indicates that there are statistical differences, and that the effect size is medium ($0.3 \leq r < 0.5$). A - signals scenarios where the algorithm in the row is worst than the one in the column. Finally, a $\sim$ indicates that no statistical differences between the algorithms were found. The statistical results confirm that evolved strategies tend to perform better in situations resembling those found during learning. Selection methods $R11 - R14$ excel in the generational scenario and one specific strategy ($R13$) is able to outperform all hand-designed approaches. When the steady-state EA surrogate is adopted, the effectiveness of the $R21 - R24$ and $R31 - R34$ strategies is evident. The

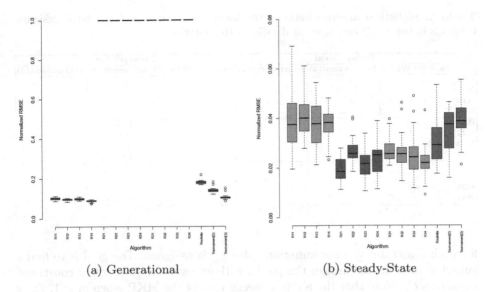

(a) Generational   (b) Steady-State

Fig. 4. MKP optimization results of the 15 selection strategies chosen for the generalization study: panels (a), (b), present the results obtained with the generational and steady state EAs, respectively.

Table 4. Statistical analysis between the learned strategies and the hand-designed methods in the KP (see text for details on the notation).

	Generational			Steady-State		
	Roulette Wheel	Tournament(2)	Tournament(3)	Roulette Wheel	Tournament(2)	Tournament(3)
R11	+++	+++	-	-	~	~
R12	+++	+++	-	-	~	~
R13	+++	+++	++	-	-	~
R14	+++	+++	-	-	-	~
R21	-	-	-	+++	+++	+++
R22	-	-	-	+++	+++	+++
R23	-	-	-	+++	+++	+++
R24	-	-	-	++	+++	+++
R31	+++	-	-	~	~	+++
R32	-	-	-	~	+++	+++
R33	+++	-	-	~	+++	+++
R34	+++	-	-	~	~	+++

performance of methods evolved with the $R2$ setting is particularly impressive, as each one of them outperforms all hand-designed selection mechanisms.

4.1   Generalization

To complete our analysis we briefly investigate if the evolved strategies generalize well to a problem different from that used in the learning step. We maintain our focus on the KP class, but consider the Multiple Knapsack Problem (MKP) variant. The MKP can be described as follows: given two sets of n items and m knapsack constraints (resources), for each item j, a profit p_j is assigned, and

Table 5. Statistical analysis between the learned strategies and the hand-designed methods in the MKP (see text for details on the notation).

	Generational			Steady-State		
	Roulette Wheel	Tournament(2)	Tournament(3)	Roulette Wheel	Tournament(2)	Tournament(3)
R11	+++	+++	~	~	~	~
R12	+++	+++	+++	-	~	~
R13	+++	+++	++	-	~	~
R14	+++	+++	+++	-	~	~
R21	-	-	-	+++	+++	+++
R22	-	-	-	~	++	+++
R23	-	-	-	+++	+++	+++
R24	-	-	-	++	+++	+++
R31	-	-	-	~	++	+++
R32	-	-	-	~	+++	+++
R33	-	-	-	~	+++	+++
R34	-	-	-	+++	+++	+++

for each constraint i, a consumption value r_{ij} is assigned. The goal is to find a subset items that maximizes the profit, without exceeding the given constraint capacities C_i. Note that the KP is a special case of the MKP when $m = 1$. For a formal definition and additional information on the MKP, please refer to [4] or [9]. For our experimental analysis we selected several MKP instances from the *OR-Library*[1]. Due to space constraints we present results obtained with a single MKP instance with $n = 250$ items and $m = 5$ constraints. However, results obtained with other instances follow the same trend.

We maintain the 15 selection strategies adopted in the previous validation analysis and keep all other optimization conditions, including the two same surrogate EAs. Figure 4 depicts the MBF box plot distribution of the selection methods, both for the generational (panel (a)) and steady-state (panel (b)) surrogates. In Table 5 we summarize the statistical comparison between the strategies considered in the generalization study. The analysis of the results reveals the exact same trend that was identified in the previous validation. Considering the performance of the evolved selection strategies, there is a clear correlation between the conditions found in the off-line learning step and those of the validation/-generalization experiments. Additionally, optimization results are competitive with those achieved by hand-designed approaches: $R11 - R14$ methods tend to outperform standard selection strategies in generational environments, whereas $R21 - R24$ and $R31 - R34$ excel in steady-state surrogates. These outcomes confirm that the GE-based HH was able to learn strategies that generalize well to different KP variants.

5   Conclusions

In this paper we proposed a GE-based HH to discover effective selection strategies for EAs. The proposed grammar is composed by symbols that allow the creation of rank-based selection strategies. We demonstrated the validity of the

---

[1] http://people.brunel.ac.uk/~mastjjb/jeb/orlib/mknapinfo.html

approach in the domain of different Knapsack Problem variants. Results obtained show that the HH framework adapts the selective pressure of the evolved mechanism, taking into account the specific features of the adopted surrogate. Despite the simplicity of the proposed grammar, the HH was able to learn effective selection strategies, competitive with standard hand-designed mechanisms regularly adopted in the literature. Moreover, the evolved strategies generalize well to different variants of the problem considered in our study.

There are several possible extensions to the work described in this paper. One possibility is to expand this framework to different problems and verify if strategies evolved in one specific optimization situation generalize well to different, possibly related, problems. Another possibility is to consider different learning design options, such as performing multiple runs to evaluate a solution, or the adoption of multiple training instances.

We will also consider several extensions to the grammar, by adding new symbols that take into account different features of the individuals belonging to the population (e.g., age).

Acknowledgments. This work was partially supported by Fundação para a Ciência e Tecnologia (FCT), Portugal, under the grant SFRH/BD/79649/2011.

References

1. Burke, E., Hyde, M., Kendall, G., Ochoa, G., Ozcan, E., Woodward, J.: A classification of hyper-heuristic approaches. In: Gendreau, M., Potvin, J.Y. (eds.) Handbook of Metaheuristics, International Series in Operations Research and Management Science, vol. 146, pp. 449–468. Springer, US (2010)
2. Group, N.W.: Augmented BNF for syntax specifications: ABNF (1 2008). ftp://ftp.rfc-editor.org/in-notes/std/std68.txt
3. Keller, R., Poli, R.: Linear genetic programming of parsimonious metaheuristics. In: IEEE Congress on Evolutionary Computation, 2007, CEC 2007, pp. 4508–4515 (2007)
4. Kellerer, H., Pferschy, U., Pisinger, D.: Knapsack Problems. Springer, Berlin (2004)
5. Lourenço, N., Pereira, F., Costa, E.: Evolving evolutionary algorithms. In: Proceedings of the Fourteenth International Conference on Genetic and Evolutionary Computation Conference Companion, GECCO '12, pp. 51–58. ACM, New York (2012)
6. Michalewicz, Z.: Genetic Algorithms + Data Structures = Evolution Programs, 3rd edn. Springer, London (1996)
7. O'Neill, M., Ryan, C.: Grammatical Evolution: Evolutionary Automatic Programming in an Arbitrary Language. Kluwer Academic Publishers, Norwell (2003)
8. Pappa, G.L., Freitas, A.: Automating the Design of Data Mining Algorithms: An Evolutionary Computation Approach, 1st edn. Springer, Berlin (2009)
9. Raidl, G.R., Gottlieb, J.: Empirical analysis of locality, heritability and heuristic bias in evolutionary algorithms: A case study for the multidimensional knapsack problem. Evol. Comput. **13**(4), 441–475 (2005)
10. Ross, P.: Hyper-heuristics. In: Burke, E.K., Kendall, G. (eds.) Search Methodologies: Introductory Tutorials in Optimization and Decision Support Techniques, Chap. 17, pp. 529–556. Springer, US (2005)

11. Tavares, J., Machado, P., Cardoso, A., Pereira, F.B., Costa, E.: On the evolution of evolutionary algorithms. In: Keijzer, M., O'Reilly, U.-M., Lucas, S., Costa, E., Soule, T. (eds.) EuroGP 2004. LNCS, vol. 3003, pp. 389–398. Springer, Heidelberg (2004)
12. Tavares, J., Pereira, F.B.: Automatic design of ant algorithms with grammatical evolution. In: Proceedings of the 15th European Conference on Genetic Programming (2012)
13. Woodward, J., Swan, J.: Automatically designing selection heuristics. In: Proceedings of the 13th Annual Conference Companion on Genetic and Evolutionary Computation, GECCO '11, pp. 583–590. ACM, New York (2011)

Interactive Evolution

Balancing User Interaction and Control in BNSL

Alberto Tonda[1], Andre Spritzer[2,3], and Evelyne Lutton[1(✉)]

[1] INRA UMR 782 GMPA, 1 Av. Brétignières, 78850 Thiverval-Grignon, France
{alberto.tonda,evelyne.lutton}@grignon.inra.fr
[2] INRIA, AVIZ Team, Bat. 660, Université Paris-Sud, 91405 ORSAY Cedex, France
[3] Universidade Federal Do Rio Grande Do Sul,
Av. Paulo Gama, 110 - Bairro Farroupilha, Porto Alegre, RS, Brazil
spritzer@inf.ufrgs.br

Abstract. In this paper we present a study based on an evolutionary framework to explore what would be a reasonable compromise between interaction and automated optimisation in finding possible solutions for a complex problem, namely the learning of Bayesian network structures, an NP-hard problem where user knowledge can be crucial to distinguish among solutions of equal fitness but very different physical meaning. Even though several classes of complex problems can be effectively tackled with Evolutionary Computation, most possess qualities that are difficult to directly encode in the *fitness function* or in the individual's *genotype description*. Expert knowledge can sometimes be used to integrate the missing information, but new challenges arise when searching for the best way to access it: full human interaction can lead to the well-known problem of user-fatigue, while a completely automated evolutionary process can miss important contributions by the expert. For our study, we developed a GUI-based prototype application that lets an expert user guide the evolution of a network by alternating between fully-interactive and completely automatic steps. Preliminary user tests were able to show that despite still requiring some improvements with regards to its efficiency, the proposed approach indeed achieves its goal of delivering satisfying results for an expert user.

Keywords: Interaction · Memetic algorithms · Evolutionary algorithms · Local optimisation · Bayesian Networks · Model learning

1 Introduction

Efficiently using algorithmic solvers to address real world problems initially requires dealing with the difficult issue of designing an adequate optimisation landscape - that is, defining the search space and the function to be optimized. The Bayesian Network Structure Learning (BNSL) problem is a good example of a complex optimisation task in which expert knowledge is of crucial importance in the formulation of the problem, being as essential as the availability

© Springer International Publishing Switzerland 2014
P. Legrand et al. (Eds.): EA 2013, LNCS 8752, pp. 211–223, 2014.
DOI: 10.1007/978-3-319-11683-9_17

of a large enough experimental dataset. By its very nature, BNSL is also at least bi-objective: its aim is to optimize the tailoring of a model to the data while keeping its complexity low. The balance between the multiple objectives has to be decided by an expert user, either *a priori* or *a posteriori*, depending on whether a mono or multi-objective solver is used. Other high level design choices made by the expert condition the type of model that is searched (i.e., the definition of the search space), and the constraints that are applied to the search.

Lack of experimental data is a rather common issue in real world instances of the BNSL problem, making the optimisation task very multi-modal or even badly conditioned. Although previous work has proved that EA approaches tend to be more robust to data sparseness than other learning algorithms [30], an efficient and versatile way of collecting expert knowledge would still represent an important progress. Interaction with the expert, for instance, can be useful to disambiguate solutions considered as equivalent given the available dataset. How to best access an expert's knowledge, however, is still an open issue: asking a human user for input at a high frequency may lead to the well-known problem of user fatigue; not asking frequently enough might result in too little feedback. In this paper we present a study that constitutes a first step into reaching this balance between interaction and automation.

For our study we developed a prototype application that allows an expert user to guide the evolution of a Bayesian network. The prototype works by alternating steps of interactive visualisation with fully automated evolution. The original network and evolved solutions are always displayed to the user as interactive node-link diagrams through which constraints can be added so that the function to be optimized can be refined. Our approach is related to humanized computation as defined by [1] (EvoINTERACTION Workshops) i.e., "systems where human and computational intelligence cooperate."

The use of interactive evolution (IEAs, or IEC) algorithms is the most common approach for humanized computation. This strategy considers the user as the provider of a fitness function (or as a part of it) inside an evolutionary loop and has been applied to various domains, such as art, industrial design, the tuning of ear implants, and data retrieval [26,28]. There are, however, different ways to interlace human interaction and optimization computations that may be as simple as what we study in this paper (i.e., an iterative scheme) or as sophisticated as collaborative learning and problem solving using Serious Games or Crowd Sourcing [4,24,31]. An interesting feature of theses latter approaches is that they consider various tools to deal with what they call "user engagement," which may represent a new source of inspiration to address the well-known "user fatigue" issue of IEAs.

This paper is organized as follows. Section 2 gives a short background on Bayesian Networks (BN) and how they can be visualized, as well as on methods used for dealing with the BNSL problem. Section 3 details our proposed approach. Experimental results are presented in Sect. 4 and an analysis is developed in Sect. 5. Finally, our conclusions and some possible directions for future research are discussed in Sect. 6.

2   Background

2.1   Bayesian Networks

Formally, a Bayesian network is a directed acyclic graph (DAG) whose nodes represent variables, and whose arcs encode conditional dependencies between the variables. This graph is called the *structure* of the network and the nodes containing probabilistic information are called the *parameters* of the network. Figure 1 reports an example of a Bayesian network.

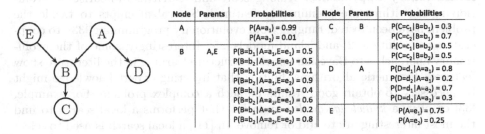

Fig. 1. Left, a directed acyclic graph. Right, the parameters it is associated with. Together they form a Bayesian network BN whose joint probability distribution is $P(BN) = P(A)P(B|A, E)P(C|B)P(D|A)P(E)$.

The set of parent nodes of a node X_i is denoted by $pa(X_i)$. In a Bayesian network, the joint probability distribution of the node values can be written as the product of the local probability distribution of each node and its parents:

$$P(X_1, X_2, ..., X_n) = \prod_{i=1}^{n} P(X_i|pa(X_i))$$

2.2   The Structure Learning Problem

Learning the optimal structure of a Bayesian network starting from a dataset is proven to be an NP-hard problem [7]. Even obtaining good approximations is extremely difficult, since compromises between the representativeness of the model and its complexity must be found. The algorithmic approaches devised to solve this problem can be divided into two main branches: heuristic algorithms (which often rely upon statistical considerations on the learning set) and score-and-search meta-heuristics. Recently, hybrid techniques have been shown to produce promising results.

Heuristic algorithms: The machine learning community features several state-of-the-art heuristics algorithms to build Bayesian network structures from data.

Some of them rely upon the evaluation of conditional independence between variables, while others are similar to score-and-search approaches, only performed in a local area of the solutions' space, determined through heuristic considerations. The main strength of these techniques is their ability of returning high-quality results in a time which is negligible when compared to meta-heuristics.

Two of the best algorithms in this category are *Greedy Thick Thinning* (GTT) [5] and *Bayesian Search* (BS) [9]. Although a detailed description of the two procedures is outside the scope of this work, it is important to highlight the most relevant difference between them. While GTT is fully deterministic, always returning the same solution for the same input, BS is stochastic, starting from different random positions at each execution. Both GTT and BS implementations can be found in commercial products such as GeNie/SMILE [12].

Evolutionary approaches: Among score-and-search meta-heuristics, evolutionary algorithms are prominently featured. Several attempts to tackle the problem have been tested, ranging from evolutionary programming [33], to cooperative co-evolution [2] and island models [25]. Interestingly, some of the evolutionary approaches to Bayesian network structure learning in the literature show features of memetic algorithms, hinting that injecting expert knowledge might be necessary to obtain good results on such a complex problem. For example, [33] employs a *knowledge-guided mutation* that performs a local search to find the most interesting arc to add or remove. In [11], a local search is used to select the best way to break a loop in a non-valid individual. The K2GA algorithm [20], in its turn, exploits a genetic algorithm to navigate the space of possible node orderings, and then runs the greedy local optimisation K2, which quickly converges on good structures starting from a given sorting of the variables in the problem.

Memetic algorithms: Memetic algorithms are *"population-based meta-heuristics composed of an evolutionary framework and a set of local search algorithms which are activated within the generation cycle of the external framework"* [18]. First presented in [23], they gained increasing popularity in the last few years [21]. What makes these stochastic optimisation techniques attractive is their ability to quickly find high-quality results while still maintaining the exploration potential of a classic evolutionary algorithm. Their effectiveness has been proven in several real-world problems [15,22] and there have been initial attempts to employ them in the structure learning problem. In particular, in [29] the authors combine the exploratory power of an evolutionary algorithm with the efficient exploitation of GTT, obtaining Bayesian network structures with higher representation and lower complexity than results produced by the most prominently featured heuristic methods.

2.3   Visualizing Bayesian Networks

It has been shown that efficient interactions in humanized computation requires efficient visualisations [19]. Current visualisation tools for BN rely on classical graph layouts for the qualitative part of the BN, i.e., its graphical structure.

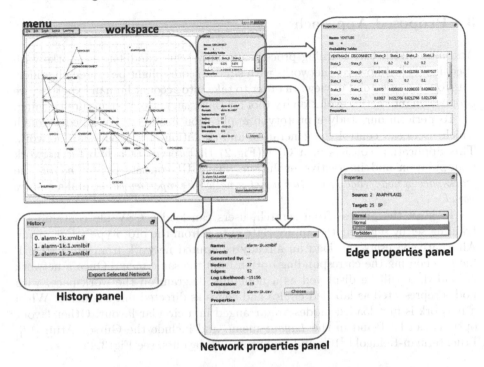

Fig. 2. Overview of the prototype's interface in use: a network being displayed and prepared for evolution. **Node properties panel:** The table shows the parameters or, in other words, the conditional probabilities for the corresponding variable. **Edge properties panel:** The arcs can be set as forced or forbidden before running the structure learning algorithms. **Network properties panel:** The log-likelihood expresses how well the current network expresses the dataset, while the dimension is a measure of the network's complexity. **History panel:** Every time a structure learning algorithm is run, a new network is added to the history.

Still, a difficult issue remains regarding the quantitative part of the BN: the conditional probability set associated to each node of the graph. It has been noted in 2005 that "the work performed on causal relation visualisation has been surprisingly low" [6]. Various solutions have been proposed like in [10], *BayViz* [6,10] *SMILE* and *GeNIe* [13], or *VisualBayes* [32]. To our knowledge, the most advanced and versatile visualisation interface for dealing with structure learning is *GeNIe*, a development environment for building graphical decision-theoretic models from the Decision Systems Laboratory of the University of Pittsburgh: it has gained a notoriety in teaching, research and industry.

None of these tools, however, has really been designed to run a smooth interaction scheme and to easily allow users to revisit the learning stage after the visualisation. Our approach explores new features for visualisation-based interactive structure learning strategies. For the moment, it does not address quantitative visualisation, though that may be considered in the future.

3   Proposed Approach

Automated structure learning processes usually score candidate networks with specific metrics: however, networks with similar scores might be extremely different from a user's point of view. In order to take into account human expertise, we propose an interactive evolutionary tool for Bayesian network structure learning.

To perform our study a prototype application has been developed through which users can control the generation and evolution of the Bayesian network. This application consists of a GUI (Fig. 2) that serves as a hub for network manipulation and interactive evolution. The GUI consists of the *menu*, the *workspace*, a *node/edge properties panel*, a *network properties panel*, and a *history panel*.

To start the process from scratch, users can load a CSV file containing a training set by selecting the appropriate option from the prototype's *File* menu. Alternatively, users can load an already computed network from an XMLBIF file by choosing the corresponding option from the same menu. Once a network is loaded, it will be displayed as a node-link diagram on the workspace, with nodes represented as labelled circles and edges as directed line segments. When a network is first loaded, nodes are arranged in a circular layout. Other layout options can be found in the *Layout* menu, and include the Gürsoy-Atun [17], Fruchterman-Reingold [16], and Sugiyama [27] layouts, see Fig. 3.

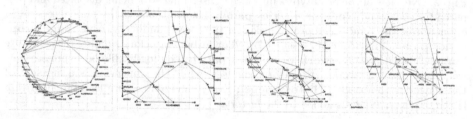

Fig. 3. Sample of layout options, from left to right: circular, Gürsoy-Atun, Fruchterman-Reingold and Sugiyama layouts of the Alarm BN benchmark.

Navigation in the workspace consists of zooming and panning. Users can zoom in or out by spinning the mouse wheel and pan using the scrollbars that appear when the visualisation is too big to fit in the workspace's view. Panning can also be performed with the *drag tool*, accessible from the *Edit* menu. When this tool is active, panning can be performed by clicking and dragging anywhere on the workspace.

By default, when a network is first loaded the *selection tool* is active. This tool allows users to select nodes and edges and move them around the workspace by clicking and dragging. Multiple objects can be selected by clicking on each object separately while the *Ctrl* key is pressed or by clicking on an empty area of the workspace and dragging so that the shown selected area intersects or covers the desired objects. Clicking and dragging on any selected object will move all others along with it.

Users can connect nodes to one another with the *Create Edge* tool, available from the *Graph* menu. Once this tool is active, the new edge can be created by first clicking on the desired origin node and subsequently on the target one. While the new edge is being created, a dashed line is shown from the origin node to the current cursor position to help users keep track of the operation. If after choosing the origin node they click on empty space instead of on another node, the edge creation is cancelled. To delete an edge from the graph, after selecting it they can either press the *Delete* key on the keyboard or select *Remove Edge* from the *Graph* menu. This operation is irreversible so a dialogue will pop up to ask for their confirmation.

When an object is selected in the workspace, its properties are displayed in the properties panel (node properties and edge properties panels of Fig. 2). Node properties include its name and numeric id in the graph as well as its probability table (if a training set has been loaded) and a list of other properties that might be present in the network's corresponding file. Edge properties show the id and name of an edge's origin and target nodes and helps users prepare the network for evolution of the network by setting the edge as *forced* or *forbidden*, or leaving it as a normal edge. Forced edges will appear in green in the workspace, while forbidden edges will appear in red.

From the moment the network is loaded, its properties are displayed in the network properties panel (Fig. 2). These properties include the amount of nodes and edges, the network's log likelihood and dimension, and other properties loaded from the network file, all updated every time there is a change in the graph. If the network was generated by evolving another, the parent network and the method used to generate it will also be shown. The training set that will be used to evolve the network can also be set from within this panel through the corresponding field's *Choose* button, which lets users load a CSV file. Note that the training set must be compatible with the network (i.e., have the exact same nodes).

If the current network has been created directly from a training set or one has been loaded in the network properties panel, it can be evolved into new networks. This is done through the learning algorithms accessible through the *Learning* menu. Users can choose among three techniques: Greedy Thick Thinning, Bayesian Search and μGP. When one is chosen, its corresponding configuration dialog is shown, where parameters for the evolution can be set and, for the case of μGP, stop conditions defined.

After evolution, the workspace is updated to display the new network. The new network is also added to the list in the history panel (Fig. 2). In this panel, the current network is always shown highlighted. Users can change the currently displayed network by clicking on its name and export it to an XMLBIF file through the *Export Selected Network* button. The latest layout is always kept when alternating among the different networks.

The prototype application was implemented in C++ using the Qt 4.8.2 framework and the Boost (http://www.boost.org) and OGDF [8] libraries. Figure 2 shows the prototype in use. A couple of networks have been generated using the learning algorithms, with the one displayed on the workspace having been created with Greedy Thick Thinning. The user has set some of the edges to forced (MINVOLSET to VENTMACH and MINVOLSET to DISCONNECT) and forbidden (INTUBATION to SHUNT) and a node has been selected (DISCONNECT).

4   Experimental Setup

In order to validate the proposed approach, test runs were performed in cooperation with two experts on food processing and agriculture. Agri-food research lines exploit Bayesian network models to represent complex industrial processes for food production.

The first expert analysed a dataset on cheese ripening [3]. It consists of 27 variables evaluating different properties of the cheese from the point of view of the producer. Of these variables, 7 are qualitative while the other 20 refer to chemical processes. A candidate solution for the dataset is reported in Fig. 4.

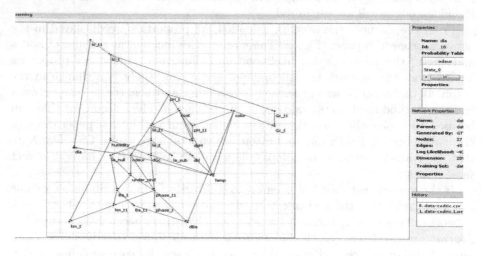

Fig. 4. A sample configuration of the complete network used in the test trial. The Sugyiama layout is preferred by the expert to visualize the structure.

The second expert analysed a dataset on biscuit baking. It consists of 10 variables describing both properties of the biscuits, such as weight and colour, and controlling variables of the process, such as heat in the top and bottom parts of the oven.

After a preliminary run, the setup of the memetic algorithm is changed in order to better fit the user's preferences. In particular, since the prototype is not optimized with regards to the running speed of the evolutionary process, the population size is reduced in comparison to the parameters reported in [30] so that a compromise can be reached between the quality of the results and time the user needs to wait before seeing the outcome.

5   Analysis and Perspectives

The expert users' response to the prototype's graphical user interface was generally positive. The ease of arc manipulation, which made it possible to immediately see the improvements in the network's representativeness and/or dimension, was well received. Also commended were the automatic layout algorithms, which were extensively used when considering the entire network. The possibility of rapidly browsing through the history of networks was used thoroughly by the experts and found to be advantageous. They felt, however, that comparing candidates would have been more immediate and effective if the interface would allow such candidates to be shown side-by-side, two at a time.

Since the process of structure learning is interactive, the users also noted how the possibility of cumulating constraints would be beneficial. In the current framework, the forced and forbidden arcs are clearly visible in each network, but they have to be set again every time a learning method is run. Despite results of slightly higher quality provided by the memetic approach, both users felt that the improvement in quality did not justify the extra time needed to obtain the solution (this approach can take up to several minutes, while the others finish running after a few seconds). For this reason, the experts favoured a more interactive approach, running the deterministic heuristic (GTT), changing the forced and forbidden arcs in its results, and repeating the process until a satisfactory solution was found.

Concerning algorithm performance, it should be noted that in order to understand the efficacy of the tool one of the users repeatedly divided the original network in smaller networks, being more confident that in this way he could highlight links that he deemed right or wrong (see Fig. 5 for an example). In networks with a reduced number of variables, however, the difference in performance between the methods became less clear, since smaller search spaces inevitably favours the heuristics. Nevertheless, the second expert was able to use the tool to eventually exclude a potential relationship between two variables in the process by iteratively generating configurations and then focusing on the log-likelihood values presented by the different candidate solutions.

Summarizing, the feedback given by the expert user in this first trial allowed us to compile a list of features that should make the structure learning experience more efficient:

1. Speeding up the memetic algorithm is recommended, and can be done straightforwardly by using parallel evaluations and letting the user tweak some internal parameters;

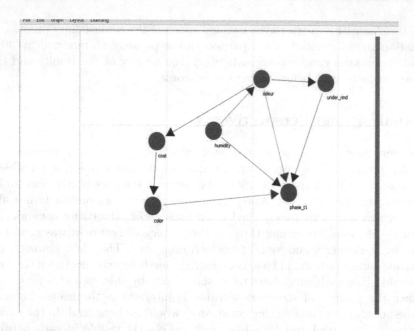

Fig. 5. One of the sub-networks extensively explored by the user. In particular, this one contains only qualitative variables from the original dataset.

2. Allowing the user to compare solutions side-by-side could be very helpful for the user, since humans are more inclined to visually compare two network at the same time than by simply browsing through the history;
3. Modifying the memetic algorithm to ask for the user's input at predetermined points (in order to try to extract his preferences by comparing networks, as in user-centric memetic algorithms [14]) might be a way to involve the user in a more time-consuming evolutionary process;
4. Designing special features to address Dynamic Bayesian Networks (DBNs). DBNs are extensively used in the agri-food field, and existing BN tools are often missing inference and learning method specifically tailored for these structures;
5. Minor features such as: allowing the user to reverse arcs; visualizing node-related statistics in pop-up windows (for clarity); selecting several arcs at the same time; and making it possible to select only a subset of variables from the original dataset.

6   Conclusion

In this paper we presented a preliminary study on balancing automatic evolution and user interaction for the NP-hard problem of Bayesian network structure learning. The study was performed through a graphical user interface.

A test run with a modelling expert showed that the tool is able to assist the user in expressing knowledge that would be difficult to encode in a classical fitness function, returning more satisfying models than a completely automatic approach. Despite the promising preliminary results, several improvements must be performed on the proposed framework to enhance usability and progress towards an optimal balance between automatic evolution of results and user interaction. For example, the evolutionary approach included at the core of the framework is found to be too time-consuming when compared to fast state-of-the-art heuristic algorithms.

Further developments will add other evolutionary structure learning algorithms, as well as the possibility for more user interaction in the definition of parameters and during the evolution itself.

Acknowledgments. The authors would like to thank Cédric Baudrit and Nathalie Perrot for contributing to this study with their expertise.

References

1. Rothlauf, F. (ed.): EvoWorkshops 2006, vol. 3907. Springer, Heidelberg (2006)
2. Barriere, O., Lutton, E., Wuillemin, P.H.: Bayesian network structure learning using cooperative coevolution. In: Genetic and Evolutionary Computation Conference, GECCO 2009 (2009)
3. Baudrit, C., Sicard, M., Wuillemin, P., Perrot, N.: Towards a global modelling of the camembert-type cheese ripening process by coupling heterogeneous knowledge with dynamic bayesian networks. J. Food Eng. **98**(3), 283–293 (2010)
4. Bellotti, F., Berta, R., De Gloria, A., Primavera, L.: Adaptive experience engine for serious games. IEEE Trans. Comput. Intell. AI Games **1**(4), 264–280 (2009)
5. Cheng, J., Bell, D.A., Liu, W.: An algorithm for bayesian belief network construction from data. In: Proceedings of AI & STAT'97, pp. 83–90 (1997)
6. Chiang, C.H., Shaughnessy, P., Livingston, G., Grinstein, G.G.: Visualizing graphical probabilistic models. Technical report 2005–017, UML CS (2005)
7. Chickering, D.M., Geiger, D., Heckerman, D.: Learning bayesian networks is NP-hard. Technical report MSR-TR-94-17, Microsoft Research, November 1994
8. Chimani, M., Gutwenger, C., Jünger, M., Klau, G.W., Klein, K., Mutzel, P.: The Open Graph Drawing Framework (OGDF). Handbook of Graph Drawing and Visualization, pp. 543–569 (2011)
9. Cooper, G.F., Herskovits, E.: A Bayesian method for the induction of probabilistic networks from data. Mach. Learn. **9**, 309–347 (1992)
10. Cossalter, M., Mengshoel, O.J., Selker, T.: Visualizing and understanding large-scale Bayesian networks. In: The AAAI-11 Workshop on Scalable Integration of Analytics and Visualization, pp. 12–21 (2011)
11. Delaplace, A., Brouard, T., Cardot, H.: Two evolutionary methods for learning bayesian network structures. In: Wang, Y., Cheung, Y., Liu, H. (eds.) CIS 2006. LNCS (LNAI), vol. 4456, pp. 288–297. Springer, Heidelberg (2007)
12. Druzdzel, M.J.: SMILE: Structural Modeling, inference, and Learning Engine and GeNIe: A Development Environment for Graphical Decision-Theoretic Models, pp. 902–903. American Association for Artificial Intelligence, Menlo Park (1999)
13. Druzdzel, M.J.: Smile: structural modeling, inference, and learning engine and genie: a development environment for graphical decision-theoretic models. In: Proceedings of AAAI'99, pp. 902–903 (1999)

14. Espinar, J., Cotta, C., Fernández-Leiva, A.J.: User-centric optimization with evolutionary and memetic systems. In: Lirkov, I., Margenov, S., Waśniewski, J. (eds.) LSSC 2011. LNCS, vol. 7116, pp. 214–221. Springer, Heidelberg (2012)
15. Fan, X.F., et al.: A direct first principles study on the structure and electronic properties of $be_x zn_{1-x} o$. Appl. Phys. Lett. **91**(12), 121121–121123 (2007)
16. Fruchterman, T.M.J., Reingold, E.M.: Graph drawing by force-directed placement. Softw. Pract. Exper. **21**(11), 1129–1164 (1991)
17. Gürsoy, A., Atun, M.: Neighbourhood preserving load balancing: a self-organizing approach. In: Bode, A., Ludwig, T., Karl, W.C., Wismüller, R. (eds.) Euro-Par 2000. LNCS, vol. 1900, pp. 234–241. Springer, Heidelberg (2000)
18. Hart, W.E., Krasnogor, N., Smith, J.E.: Memetic evolutionary algorithms. In: Hart, W.E., Smith, J.E., Krasnogor, N. (eds.) Recent Advances in Memetic Algorithms. Studies in Fuzziness and Soft Computing, vol. 166, pp. 3–27. Springer, Heidelberg (2005)
19. Hayashida, N., Takagi, H.: Visualized IEC: interactive evolutionary computation with multidimensional data visualization. In: IECON 2000. 26th Annual Conference of the IEEE, vol. 4, pp. 2738–2743 (2000)
20. Larranaga, P., et al.: Learning bayesian network structures by searching for the best ordering with genetic algorithms. IEEE Trans. Syst. Man Cybern. Part A: Syst. Hum. **26**(4), 487–493 (1996)
21. Neri, F., Cotta, C.: Memetic algorithms and memetic computing optimization: a literature review. Swarm Evol. Comput. **2**, 1–14 (2012)
22. Nguyen, Q.H., Ong, Y.S., Lim, M.H.: A probabilistic memetic framework. IEEE Trans. Evol. Comput. **13**(3), 604–623 (2009)
23. Norman, M., Moscato, P.: A competitive and cooperative approach to complex combinatorial search. In: Proceedings of the 20th Informatics and Operations Research Meeting, pp. 3–15 (1991)
24. Potter, A., McClure, M., Sellers, K.: Mass collaboration problem solving: a new approach to wicked problems. In: 2010 International Symposium on Collaborative Technologies and Systems (CTS), pp. 398–407, May 2010
25. Regnier-Coudert, O., McCall, J.: An Island model genetic algorithm for Bayesian network structure learning. In: IEEE CEC, pp. 1–8 (2012)
26. Simons, C., Parmee, I.: User-centered, evolutionary search in conceptual software design. In: IEEE Congress on Evolutionary Computation, 2008, CEC 2008. (IEEE World Congress on Computational Intelligence), pp. 869–876, June 2008
27. Sugiyama, K., Tagawa, S., Toda, M.: Methods for Visual Understanding of Hierarchical System Structures. IEEE Trans. Syst. Man Cybern. SMC **11**(2), 109–125 (1981)
28. Takagi, H.: New topics from recent interactive evolutionary computation researches. In: Lovrek, I., Howlett, R.J., Jain, L.C. (eds.) KES 2008, Part I. LNCS (LNAI), vol. 5177, p. 14. Springer, Heidelberg (2008)
29. Tonda, A., Lutton, E., Squillero, G., Wuillemin, P.-H.: A memetic approach to bayesian network structure learning. In: Esparcia-Alcázar, A.I. (ed.) EvoApplications 2013. LNCS, vol. 7835, pp. 102–111. Springer, Heidelberg (2013)
30. Tonda, A.P., Lutton, E., Reuillon, R., Squillero, G., Wuillemin, P.-H.: Bayesian network structure learning from limited datasets through graph evolution. In: Moraglio, A., Silva, S., Krawiec, K., Machado, P., Cotta, C. (eds.) EuroGP 2012. LNCS, vol. 7244, pp. 254–265. Springer, Heidelberg (2012)
31. Voulgari, I., Komis, V.: On studying collaborative learning interactions in massively multiplayer online games. In: 2011 Third International Conference on Games and Virtual Worlds for Serious Applications (VS-GAMES), pp. 182–183, May 2011

32. Williams, L., Amant, R.S.: A visualization technique for Bayesian modeling. In: Proceedings of IUI'06 (2006)
33. Wong, M.L., Lam, W., Leung, K.S.: Using evolutionary programming and minimum description length principle for data mining of Bayesian networks. IEEE Trans. Pattern Anal. Mach. Intell. **21**(2), 174–178 (1999)

Parallel Evolutionary Algorithms

Massively Parallel Generational GA on GPGPU Applied to Power Load Profiles Determination

Frédéric Krüger(✉), Daniel Wagner, and Pierre Collet

ICUBE Laboratory, Universit de Strasbourg – France Pôle API,
Rue Sbastien Brant, 69121 Illkirch, France
frederic.kruger@etu.unistra.fr

Abstract. Evolutionary algorithms are capable of solving a wide range of different optimization problems including real world ones. The latter, however, often require a considerable amount of computational power. Parallelization over powerful GPGPU cards is a way to tackle this problem, but this remains hard to do due to their specificities. Parallelizing the fitness function only yields good results if it dwarfs the rest of the evolutionary algorithm. Otherwise, parallelization overhead and Amdahl's law may ruin this effort.

In this paper, we will show how completely parallelizing an evolutionary algorithm can help solving a large real world electrical problem with a lightweight evaluation function without quality loss.

1 Introduction

Real world optimization problems often require a considerable amount of computational power due to their high complexity. The use of GPGPU (General Purpose Graphics Processing Units) revolutionizes evolutionary computation as it allows to tackle a broader range of problems which, until now, were out of reach for standard sequential evolutionary algorithms.

Many papers show the type of speedups that can be achieved on benchmark functions as well as on real world problems (chemistry) by parallelizing the evaluation function only on a single GPGPU card [1,3,4] and GPGPU specific architecture is described in several publications [2–4].

However, this particular way to parallelize an evolutionary algorithm requires extremely time consuming fitness functions. Problems with lightweight fitness functions cannot take advantage of the GPGPUs computing power. In [2], the authors describe a complete parallel evolutionary algorithm and present speedups on benchmark functions showing that solving problems with lightweight evaluation functions can also benefit from GPGPU parallelization.

In this paper, we propose to determine load profiles using a generational evolutionary algorithm completely parallelized on a single GPGPU chip. The obtained speedups are shown and the quality of the results discussed. Moreover, the influence of parameters specific to GPGPU computing is also examined.

The authors would like to thank the financial contribution of the "Région Alsace".

© Springer International Publishing Switzerland 2014
P. Legrand et al. (Eds.): EA 2013, LNCS 8752, pp. 227–239, 2014.
DOI: 10.1007/978-3-319-11683-9_18

2   Parallel Genetic Algorithm

2.1   Generational Evolutionary Algorithm

The specificity of a generational evolutionary algorithm is that it creates a population of children that is the same size as the population of parents. Then, parents are discarded and the next generation is the children population.

A "refinement" can be added in order to implement an "elitist" generational EA: a parent is replaced by one of his children only if the latter is better than the parent. Otherwise, the rest of the algorithm is quite standard: the initial population is generated randomly and children are generated thanks to genetic variation operators applied to selected parents.

This evolutionary loop is repeated until a stopping criterion is met.

2.2   Full Parallelization of a Generational Evolutionary Algorithm

Each step of the generational algorithm is parallelized on a GPGPU card: after memory allocation of two population spaces on the device (for a parent population and an offspring population), all cores start with an initialization stage where an initial population is created and individuals are evaluated in parallel.

Because all these steps are nearly identical for all cores, divergence between cores is minimal, which is essential on an SIMD/SPMD architecture.

Then, for each generation, an evolutionary kernel is launched from the host. It assigns as many threads as there are individuals in the parent population and it dispatches the threads among the core blocks. The evolutionary kernel selects 2 parents from the parent population and performs a crossover between them to create an offspring that is stored in the offspring population space on the device. The genome of the offspring is then mutated and evaluated.

Finally, the freshly created offspring is compared to the parent with the same thread ID. If the offspring is better, it replaces the parent. If not, the parent is kept for the next generation.

Meanwhile, the host waits for all the threads on the device to synchronize, in order to launch the evolution kernel for the next generation.

The selection of the parents for reproduction is performed through a tournament selection operator.

2.3   Random Number Generation: Host API *vs.* Device API

Pseudo random number sequences can be generated in two different ways using the CURAND library [5]: by using the host API or by using the device API.

In the host API (Host_RBG), random numbers are produced by generators. The production of numbers requires the creation of a generator and memory allocation on the device. Then, random numbers are generated in parallel directly on the device, ready to be used by subsequent kernels. To maximize the efficiency of the generator, a great amount of random numbers should be generated simultaneously.

In the device API (Dev_RNG), each thread has its own generator and its own seed. The states of the generators, after being initialized, are stored in device memory. The pseudo random numbers are then generated directly on the device. The advantage of this API is that the generation function can be called from device functions while in the host API, random number are generated through a specific kernel beforehand. While being faster than the host API, the device API provides less guarantees about the mathematical properties of the generated sequences.

3   Determination of Load Profiles

Energy distribution companies such as Électricité de Strasbourg Réseaux (ESR) struggle to obtain very precise estimations of the energy demand of large scale as well as medium scale electric networks. They have access to their power load profiles that strive to approximate the behavior of specific end user classes, but these profiles are often not very precise as they do not take into account factors such as the presence of electrical heating or the type of housing. Ignoring these factors results in inaccurate estimations of load curves for areas very sensitive to temperature changes. For instance branches of the electrical network that distribute energy mostly to end users with electrical heating are very sensitive to temperature drops in the winter time.

3.1   Presentation of the Case Study

The determination of load profiles is a very prolific area of research [6–12]. There are a multitude of different methods to obtain profiles of good quality. Nevertheless, most of these methods rely on time series of end user load measurements obtained through measurement campaigns. These load measurements time series are very expensive and time consuming to acquire and hence not always available. However, a paper of 2003 [13] shows that it is possible to obtain high quality load profiles without any prior knowledge of the electrical network by considering the load profile determination problem as a blind source separation problem. The feasibility of the method is proven on artificial datasets only.

3.2   General Setup and Methodology

Available Information. The information used in the case of this real world problem is provided by Électricité de Strasbourg Réseaux. The information includes the topology of the electric network maintained by ÉSR as well as load measurements performed at different levels of the network:

1. At the level of the 20 kV HV (high voltage) feeders: average load measurements performed at a 10 min step
2. At the LV (low voltage) end user level: biannual energy meter readings from which an average load can be calculated

For the purpose of this study, the end users are separated into 8 different classes according to the following criteria:

- Usage: **domestic** or **professional**
- Type of housing: **apartments** or **single houses**
- Presence of electrical heating: **with** or **without**
- Tariff: **single rate** or **double rate**

The presence of electrical heating was determined by machine learning techniques applied on end user energy consumption history. The double rate tariff represents the peak/offpeak tariff.

Methodology. The method applied to determine the new load profiles is similar to the method presented in [13]. We assume that the load curve of a 20 kV feeder d is equal to the weighted sum of the different profiled load curves of the end users fed by d. The determination of the load profiles is hence very similar to a blind source separation problem [14]: separate a set of source signals (the set of profiles) from a set a mixed signals (the set of 20 kV feeder load curves). The problem can be summed up by the following equation:

$$\begin{pmatrix} W_{1,1} & \cdots & W_{1,8} \\ \vdots & \ddots & \vdots \\ W_{m,1} & \cdots & W_{m,8} \end{pmatrix} \times \begin{pmatrix} P_1 \\ \vdots \\ P_8 \end{pmatrix} + \begin{pmatrix} \epsilon_1 \\ \vdots \\ \epsilon_m \end{pmatrix} = \begin{pmatrix} p_1 \\ \vdots \\ p_m \end{pmatrix} \tag{1a}$$

$$W_{m,8} \times P_{8,1} + \epsilon_{m,1} = p_{m,1} \tag{1b}$$

where:

- $W_{m,8}$ the weight matrix: 8 end user average load for m 20 kV feeders;
- $P_{8,1}$ the profile matrix for the 8 end user classes (set of source signals);
- $\epsilon_{m,1}$ the residue matrix for m 20 kV feeders;
- $p_{m,1}$ the power matrix for m 20 kV feeders (set of mixed signals);

The residue matrix is necessary as the data used contains a certain amount of noise and error caused by:

- Changes in the network topology (load transfer)
- Measurement device failures
- Power loss
- Other unknown elements

The consequence of these factors is a high amount of residue in $\epsilon_{m,1}$. Therefore, it is not possible to determine the linear independence of the system, a prerequisite to a classic blind source separation. The linear independence of the system was therefore assumed for the rest of this study. A genetic algorithm (GA) was chosen to solve the blind source separation, more than adequate to tackle this class of problems [15,16], regardless of the amount of error in the data [17].

3.3 Engine of the Genetic Algorithm

Fitness Function. The fitness function is a mix of two distinct sub-functions: a function to compare the estimated load curves with the measured load curves and a function to measure the smoothness of the load profiles.

The following equation calculates for a single 20 kV feeder c the difference between the estimated load curve constructed with the profiles from the individual's genotype and the measured load curve:

$$F_c = \frac{\sum_{t_0}^{t_1} |p(t) - (W_1 P_1(t) + \ldots + W_n P_n(t))|}{p(t)} \tag{2}$$

where:

- $p(t)$ load of the feeder at time t (in kW);
- W_n average power of end users with profile n (in kW);
- $P_n(t)$ value of profile n at time t (no unit);

The second equation determines the smoothness of the profiles:

$$Dist = \sum_{i=1}^{n} MAX((d - D_i), 0) \tag{3}$$

where:

- n number of profiles (8 in the case of this study);
- d overall average distance between two half-hour points in a profiles;
- D_i average distance between two half-hour points of profile i;

The total fitness F is determined over the complete set of m measured load curves by combining the two equations presented above:

$$F = \frac{\sum_{c=1}^{m} F_c + \sum_{i=1}^{n} MAX((d - D_i), 0)}{n + m} \tag{4}$$

where:

- m number of measured load curves in the mixed signals set;

The goal of the genetic algorithm is to find the individual $i.e.$ the set of profiles that minimizes the fitness value F that is very simple to compute (only simple additions and a couple of divisions), making for a very light evaluation function.

Individual genome. An individual represents a set of 8 profiles for a specific day of a given month. Each profile is represented as an array of 48 floating points (one for every half-hour in the day). Hence, the genotype of an individual is a 48×8 long float point array.

Engine of the original genetic algorithm. High quality power load profiles (reference profiles) have already been found using an elitist genetic algorithm generated with EASEA [18]. Nevertheless, the determination of load profiles for a single day of the year requires about 13 min. The determination of load profiles for a complete year therefore requires about 79 h! The latter are not static. They are bound to evolve with end user energy consumption habits and will require regular redefinitions. Accelerating their creation would enhance the current process.

4   Results

The relevance and the quality of the load profiles found by the GA have been presented in another paper [18] and will not be discussed in this paper. Instead, the different speedups that can be achieved by parallelizing the problem on a single GPGPU card are observed and the benefit of using GPGPUs for this particular problem will be presented. All the values exposed in this section represent an average over 30 runs.

The experiments were performed on:

- 1 core of an Intel Core i7 CPU running at 3.33 GHz
- 512 cores of one nVidia GTX 590 processor.

The processors of an nVidia GTX 590 have 16 multi processors with 32 cuda cores. The maximum number of thread per block is 1024 and the maximum number of registers per block is 32768. Each processor of a GTX 590 has access to 1,49 GB of global memory.

The fitness function presented above requires 42 registers per thread. Therefore, the maximum number of threads per block is 780 as those threads take 32760 of the 32768 registers of the block. For security reasons, the maximum number of threads per block for this function was capped at 512. Maximum occupancy for one card is reached for $16 \times 512 = 8192$ threads.

The GA manages different data along with set of feeder load curves:

1. 2 populations: a parent population and an offspring population ($2 \times popSize \times$ 8 $times$48 values)
2. 1 vector with the mutation probability of each gene of every individual (vector size: $popSize \times 8 \times 48$)
3. 1 vector with the mutation values of each gene of every individual (vector size: $popSize \times 8 \times 48$)
4. 1 vector with the barycentric crossover weight of every individual (vector size: $popSize$)
5. 1 vector with the fitness value of every individual (vector size: $popSize$)

The total number of real number values required by the GA is $popSize \times 2306$. It is important to note that the arrays with mutation probabilities, mutation values and crossover weights are only used in the case Host_RNG.

Table 1 shows that for population sizes above 131072, the GPGPU card is out of memory. The maximum population size that can be used for this particular problem is of 131072 individuals.

Table 1. Memory in GB required by the parallel GA on GPGPU with regards to population size. Total global memory on a single GTX590 card: 1,49 GB

Population size	Total amount of memory required by the GA (in GB)
64	0,0004
1024	0,007
8192	0,05
32768	0,23
131072	0,93
262144	1,87

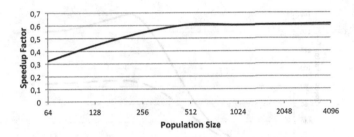

Fig. 1. Speedup for **CpuEzElGa** *vs.* **GpuEzElGa**

Terminology. In order to make the following section more understandable, the different algorithms compared during the determination of speedups are described and given a specific name used in the following section:

1. **CpuEzElGa:** Elitist (*El*) genetic algorithm (*Ga*) generated with EASEA (*Ez*) running on CPU (*Cpu*)
2. **GpuEzElGa:** Elitist (*El*) genetic algorithm (*Ga*) generated with EASEA (*Ez*) running on GPU (*Gpu*)
3. **GpuGenGaWiPt:** Generational (*Gen*) genetic algorithm (*Ga*) with generational population transfer (*WiPt*) running on GPU (*Gpu*)
4. **GpuGenGaWoPt:** Generational (*Gen*) genetic algorithm (*Ga*) without generational population transfer (*WoPt*) running on GPU (*Gpu*)

In some cases, a generational information transfer is performed from the GPU to the CPU. This information transfer is performed in order to compute and display statistics relative to the population convergence such as best fitness, average fitness etc... While this information transfer is not relevant when the algorithm is already tuned, it is vital while during engine parameters optimization. Hence it is only fair to compare the speedups for both configurations.

4.1   Speedups

Figure 1 shows the speedup obtained by using the automatic EASEA GPU parallelization feature: only the fitness function is parallelized on a single GPGPU card. Maximum "speedup" of 0.6 is achieved for population sizes starting at 512 individuals. The values presented in Fig. 1 are hardly speedups and can be referred to as "speeddowns". As a matter of fact, Fig. 1 shows that the genetic algorithm running on the CPU is faster than the GPU version generated with EASEA. Figure 1 shows that the fitness function of this real world problem is not intense enough computationally time-wise in order to benefit from GPU parallelization.

No comparisons have been performed for population sizes greater than 4096 individuals due to the ridiculously enormous computation times.

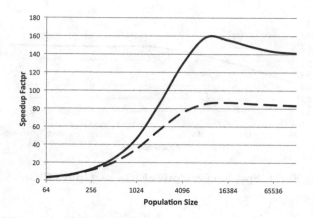

Fig. 2. Speedup for **CpuEzElGa** *vs.* **GpuGenGaWiPt** (dashed curve) and **CpuEzElGa** *vs.* **GpuGenGaWoPt** (lined curve)

Figure 2 shows the speedup obtained by parallelizing the generational genetic algorithm completely on the GPGPU card. The dashed curve shows the speedup obtained when a generational population transfer is performed, the lined curve shows the speedup obtained when **no** generational population transfer is performed.

The same figure shows that a speedup of ×160 is achieved for **GpuGen-GaWoPt** whereas a speedup of "only" ×90 is achieved for **GpuGenGaWiPt**. A speedup drop can be noticed for population sizes greater than 8192 individuals. These drops are due to the GPU card having reached maximum occupancy (both thread and registry wise) as well as to the rising computation time for random number generation/transfer.

4.2   Convergence Comparison

The second aspect to be observed is the differences in convergence for different configurations.

In order to use the GPGPU to speedup the profile determination problem, the engine of the genetic algorithm needs to be transformed from an elitist engine

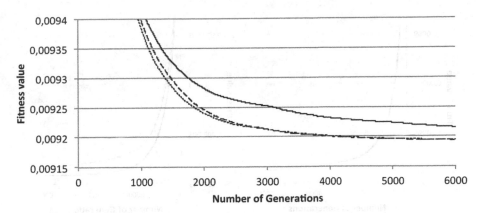

Fig. 3. Convergence comparison: elitist GA running on CPU (dotted curve), generational GA running on CPU (dashed curve) and generational GA running on GPU (lined curve).

to a generational engine. This modification has an impact on the convergence and it is of the upmost importance to make sure that this impact is minimal.

Figure 3 shows the convergence of best individual fitness for the elitist GA running on CPU (dotted curve), the generational GA running on CPU (dashed curve) and the generational GA running on GPU (lined curve). Figure 3 demonstrates that the change of evolutionary engine does not impact the way the GA converges. Moreover, the transfer of the GA from the CPU to the GPU does not seem to have a significant impact on the convergence either. Nevertheless, the impact of the visible difference in the final fitness values on the shape of the resulting profiles has to be quantified.

4.3   Influence of Different Factors

When using GPGPU programming, several features can be tweaked in order to improve speedup performances as well as result precision. Speedups can be improved by using Dev_RNG. Result quality can be improved by using double precision, which, however, restricts the maximum population size.

The left chart of Fig. 4 shows the influence of Host_RNG (lined curve) over Dev_RNG (dashed curve). This chart clearly reveals that the GA using Host_RNG reaches a far better final fitness value compared to the GA using Dev_RNG. That result is not surprising considering that Dev_RNG provides less guarantees about he mathematical properties of the generated sequences. In the case of this particular problem, only Host_RNG should be used.

The right chart of Fig. 4 shows the influence of double precision (dashed curve) over single precision (lined curve). Single precision reaches better results than double precision.

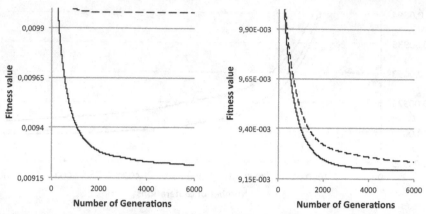

Influence of the **host** random number generation API (lined curve) and the **device** random number generation API (dashed curve) on the convergence of the GA.

Influence of **double** **precision** (dashed curve) over **single precision** (lined curve) on the convergence of the GA.

Fig. 4. Influence of different factors on the convergence of the GA.

4.4 Profile Quality Comparison

For real world problems, the quality of the results can be more important than the time required to obtain it. In this case, it is very important to make sure that the profiles obtained with the different GA configurations are equivalent to the reference profiles found by CpuEzElGa.

Figure 5 presents a bar chart comparing the correlation coefficients between the profiles found by CpuEzElGa and the profiles found by:

- GpuGenGa with a speedup optimal population size of 8192 individuals using Host_RNG (in black)
- GpuGenGa with the original optimal population size of 750 individuals using Host_RNG (in dark gray)
- GpuGenGa with the original optimal population size of 750 individuals using Dev_RNG (in light gray)

Figure 5 shows that the profiles found with the parallel GAs using Host_RNG have close to 1 correlation coefficients. On the other hand, the parallel GA using Dev_RNG performs very badly as the profiles have an average correlation coefficient of 0.5! Figure 5 also reveals that using a greater population size does not impact the quality of the results.

4.5 Discussion

Several conclusions can be drawn from the results presented above. Maximum speedup of ×160 is achieved for this particular problem with a population size

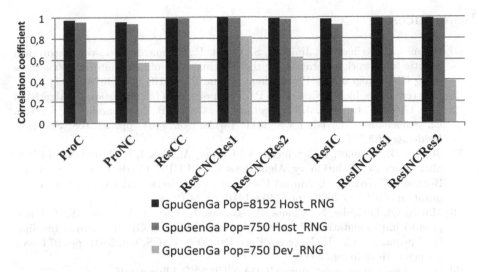

Fig. 5. Correlation coefficient between the reference profiles obtained with the elitist genetic algorithm on CPU and the profiles obtained through different GA configurations running on GPU.

of 8192. However, the population size of the original elitist genetic algorithm is only of 750 individuals and Fig. 5 shows that the use of a greater population size does not improve result quality. Therefore, if the time required by the original GA is compared to the time required by the generational GA running on GPU with a population size of 8192, a speedup of ×14 is achieved. Figure 2 shows that for a population size of 750 individuals, a speedup of ×34 is achieved. It is therefore more interesting to conserve the original population size even if it means not using the GPU card to its full capacity.

The profile determination for a whole year which requires 79 h on CPU only takes about 2 h when it is ported on GPGPU!

5   Conclusion

This paper shows how a genetic algorithm completely parallelized on GPGPU can contribute in solving efficiently and little computation time a complex problem such as power load profile determination. The speedups presented in this paper range between ×90 and ×160, reducing the computation time from 79 h to a little more than 2 h. Different aspects of GPGPU computation have been tested and their influence on GA convergence as well as profile quality examined.

The parallel genetic algorithm presented in this paper has not yet been integrated into the EASEA language. The integration has to be performed in order to allow people without GPGPU programming knowledge to take advantage of computation power of these cards to solve their optimization problem.

References

[1] Baumes, L., Krüger, F., Jimenez, S., Collet, P., Corma, A.: Boosting theoretical zeolitic framework generation for the determination of new materials structures using GPU programing. Phys. Chem. Chem. Phys. **13**, 4674–4678 (2011)

[2] Maitre, O., Lachiche, N., Collet, P.: Two ports of a full evolutionary algorithm onto GPGPU. In: Hao, J.-K., Legrand, P., Collet, P., Monmarché, N., Lutton, E., Schoenauer, M. (eds.) EA 2011. LNCS, vol. 7401, pp. 97–108. Springer, Heidelberg (2012)

[3] Maitre, O., Baumes, L., Lachiche, N., Corma, A., Collet, P.: Coarse Grain Parallelization of Evolutionary Algorithms on GPGPU Cards with EASEA. In: Rothlauf, F. (ed.) 11th Annual Conference on Genetic and Evolutionary Computation (GECCO 2009), pp. 1403–1410. ACM (2009)

[4] Maitre, O., Lachiche, N., Clauss, P., Baumes, L., Corma, A., Collet, P.: Efficient parallel implementation of evolutionary algorithms on GPGPU cards. In: Sips, H., Epema, D., Lin, H.-X. (eds.) Euro-Par 2009. LNCS, vol. 5704, pp. 974–985. Springer, Heidelberg (2009)

[5] cora.gridlab.univie.ac.at/docs/CUDA/CURAND_Library.pdf

[6] Jardini, J.A., Tahan, C.M., Gouvea, M.R., Ahn, S.U., Figueiredo, F.M.: Daily load profiles for residential, commercial and industrial low voltage consumers. IEEE Trans. Power Delivery **15**(1), 375–380 (2000)

[7] Gerbec, D., Gasperic, S., Smon, I., Gubina, F.: Allocation of the load profiles to consumers using probabilistic neural networks. IEEE Trans. Power Syst. **20**(2), 548–555 (2005)

[8] Chen, C.C., Wu, T.H., Lee, C.C., Tzeng, Y.M.: The application of load models of electric appliances to distribution system analysis. IEEE Trans. Power Syst. **10**(3), 1376–1382 (1995)

[9] Chen, C.S., Kang, M.S., Hwang, J.C., Huang, C.W.: Temperature effect to distribution system load profiles and feeder losses. IEEE Trans. Power Syst. **16**(4), 916–921 (2001)

[10] Espinoza, M., Joye, C., Belmans, R., De Moor, B.: Short-term load forecasting, profile identification, and customer segmentation. IEEE Trans. Power Syst. **20**(3), 1622–1630 (2005)

[11] Figueiredo, V., Rodriguez, F., Vale, Z., Gouveia, J.B.: An electric energy consumer characterization framework based on data mining techniques. IEEE Trans. Power Syst. **20**(2), 596–602 (2005)

[12] Capasso, A., Grattieri, W., Lamedica, R., Prudenzi, A.: A bottom-up approach to residential load modeling. IEEE Trans. Power Syst. **9**(2), 957–964 (1994)

[13] Liao, H., Niebur, D.: Load profile estimation in electric transmission networks using independent component analysis. IEEE Trans. Power Syst. **18**(2), 707–715 (2003)

[14] Acharyya, R.: A New Approach for Blind Source Separation of Convolutive Sources. VDM Verlag, Saarbrücken (2008)

[15] Rojas, I., Clemente, R.M., Puntonet, C.G.: Nonlinear Blind Source Separation Using Genetic Algorithms. Independent Component Analysis and Signal Separation (2001)

[16] Shyr, W.-J.: The hybrid genetic algorithm for blind signal separation. In: King, I., Wang, J., Chan, L.-W., Wang, D.L. (eds.) ICONIP 2006. LNCS, vol. 4234, pp. 954–963. Springer, Heidelberg (2006)

[17] Katou, M., Arakawa, K.: Blind source separation in noisy and reverberating environment using genetic algorithm. In: Proceeding of 2009 APSIPA Annual Summit and Conference (2009)

[18] Krüger, F., Wagner, D., Collet, P.: Using a genetic algorithm for the determination of power load profiles. In: Esparcia-Alcázar, A.I. (ed.) EvoApplications 2013. LNCS, vol. 7835, pp. 162–171. Springer, Heidelberg (2013)

Swarm Intelligence

O-BEE-COL: Optimal BEEs
for COLoring Graphs

Piero Consoli[1] and Mario Pavone[2]([✉])

[1] School of Computer Science, University of Birmingham,
Edgbaston Birmingham B15 2TT, UK
p.a.consoli@cs.bham.ac.uk
[2] Department of Mathematics and Computer Science,
University of Catania, V.le A. Doria 6, 95125 Catania, Italy
mpavone@dmi.unict.it

Abstract. Graph Coloring, one of the most challenging combinatorial problems, finds applicability in many real-world tasks. In this work we have developed a new artificial bee colony algorithm (called O-BEE-COL) for solving this problem. The special features of the proposed algorithm are (i) a SmartSwap mutation operator, (ii) an optimized GPX operator, and (iii) a temperature mechanism. Various studies are presented to show the impact factor of the three operators, their efficiency, the robustness of O-BEE-COL, and finally the competitiveness of O-BEE-COL with respect to the state-of-the-art. Inspecting all experimental results we can claim that: (a) disabling one of these operators O-BEE-COL worsens the performances in term of the Success Rate (SR), and/or best coloring found; (b) O-BEE-COL obtains comparable, and competitive results with respect to state-of-the-art algorithms for the Graph Coloring Problem.

Keywords: Swarm intelligence · Artificial bee colony · Graph coloring problem · Combinatorial optimization

1 Introduction

Graph coloring is one of the most popular and challenging combinatorial optimization problems, playing a central role in graph theory. It can be formalized as follow: given an undirected graph $G = (V, E)$ a coloring of G is a mapping $c : V \rightarrow S$ ($\subseteq \mathbb{N}^+$) that assigns a positive integer to each vertex in V such that $c(u) \neq c(v)$ if u and v are adjacent vertices. The elements in S represent the available *colors*. The optimization version of *Graph Coloring Problem* (GCP) asks to find a mapping c with $S = \{1, 2, \ldots, k\}$ being of minimal size, i.e., finding the smallest integer k such that G has a $k - coloring$. This minimal cardinality of S is known as the *chromatic number* of G ($\chi(G)$). Thus formally, if $k > \chi$ then a graph G is called $k - colorable$, otherwise G is $k - chromatic$ if $k = \chi$. Computing the chromatic number of a graph is an NP–complete problem [17].

© Springer International Publishing Switzerland 2014
P. Legrand et al. (Eds.): EA 2013, LNCS 8752, pp. 243–255, 2014.
DOI: 10.1007/978-3-319-11683-9_19

Tackling and solving the GCP becomes crucial and important since it has a natural applicability in many real-world problems, such as scheduling [26], time tabling [12], manufacturing [19], frequency assignment [16], register allocation [8] and printed circuit testing [18]. The GCP can be tackled following two different approaches: *assignment* or *partitioning*. The first approach consist in the classical assignment of colors to vertices; whilst the latter one is based on partitioning the set of vertices V into k disjoint subsets $(V_1, V_2, \ldots, V_k)$ such that in any subset no two vertices are linked by an edge, i.e. if u and v are in V_i (for some $i \in \{1, \ldots, k\}$) then $(u, v) \notin E$. Every subset V_i represents a color class and forms an *Independent Set* of vertices. Although several pure population–based algorithms have been used to tackle the GCP, a hybrid approach where local search methods, specialized operators and evolutionary algorithms (EAs) are combined [25] might be more effective. This is, of course, due to the intractable nature of the GCP [5].

In this work we propose an *Artificial Bee Colony (ABC)* [24] algorithm for the GCP, based on three main features: (1) a new mutation operator, (2) an optimized version of the Greedy Partitioning Crossover (GPX) [15], and (3) a temperature mechanism. The ABC algorithm is a rather recent optimization technique inspired by the intelligent foraging behavior of a colony of bees, whose strength lies in the collective behavior of self-organized swarms that individually behave without any supervision. During the last decade, ABC has attracted quite a number of researchers, and it has been successfully applied mainly to continuous optimization problems [3,23], whilst, rather few works have appeared concerning discrete optimization problems (see, for example, [27,31]). In many cases the results obtained by ABC, including the ones of this work, demonstrate that this metaheuristic is able to compete with, and sometimes even outperforms, existing state-of-the-art algorithms for difficult optimization problems.

2   O-BEE-COL: An Artificial Bee Colony

The ABC algorithm takes inspiration from the intelligent foraging behavior of bees from a beehive. It is based on three main components: (1) *food source position*, corresponding to a feasible solution to the given problem; (2) *amount of nectar*, which indicates the quality of the solution; and (3) the bee types: *employed*; *onlooker*; and *scouts* bees. The first ones have the purpose to search for food sources, and, just found, storing their information. The onlooker bees select, and exploit the better food sources found taking advantage of the information learned from employed bees. Once one of the food sources is exhausted, the employed bees associated with it become scout bees, with the purpose to discover new food sources. Once discovered, they become again employed bees.

A new ABC heuristic has been developed in order to effectively coloring a generic graph. This algorithm is henceforth referred to as *"Optimal BEEs for COLoring"* (O-BEE-COL). The algorithm begins with the creation of the initial population, where each bee represents a permutation of vertices. Because the choice of the starting points in the search space become crucial we have designed, and studied, three variants of O-BEE-COL in order to create the initial population. In the basic variant, it is randomly generated via a uniform distribution. The second variant, instead, uses a version partially randomize of RLF (Recursive Largest First) algorithm [10]. Of course, as expected, with this last variant O-BEE-COL shows better performances, because it begins the search from good solutions than the first variant. On the other hand, however using this second variant we have the disadvantage to get trapped into local optima easily, mainly in more complex instances. Thus, we have developed a third variant that is a mixed of the two previous ones. In this way, we introduce more diversity in the population in order to better exploration the search space, escaping from local optima, and exploiting good solutions at the same time. Analysing the comparisons among the three variants (not included in this paper due to limited space), the mixed one has produced the best performances obtaining better coloring in all instances tested. For example, if we take into account the "$le450\_15c$" DIMACS instance [22], using the first variant the algorithm starts from a best solution found of 28 colors and improves the coloring until to reach a solution with 20 colors. Instead with the randomized RLF, although O-BEE-COL begins from 24 colors as best solution, it never improves this coloring found. If, however, O-BEE-COL incorporates the mixed variant, starting from a best solution of 24 colors (the one found by randomized RLF), at the end of the evolution it is able to coloring the graph with 15 colors, which is also the chromatic number for this instance.

The strength of O-BEE-COL is based on three main operators: mutation operator called *SmartSwap*; optimized version of GPX [15]; and Temperature mechanism, as in Simulated Annealing, which has the aim of self-regulating of some parameters of the algorithm. The mutation operator tries to reduce the number of colour classes deleting one of them, and reassigning its vertices inside other classes. Albeit is reasonable to think that this process might be easily performed in the smaller class, unfortunately often belong to it the most troublesome nodes, i.e. the ones harder to be handled. Thus, SmartSwap works primarily on these troublesome nodes with the aim to replace them with the ones more easy to be handled. In this way becomes easier the reassignment of the vertices, and therefore the delete of the class. To do that, SmartSwap allows a fixed number of constrains unsatisfied, which will be removed via the crossover operator: only *partLimit* constraints unsatisfied are allowed. With this operator we attempt to avoid that the solutions get trapped into local optima. Greedy Partitioning Crossover – GPX – is a well-known crossover originally proposed in [15], and based on strategy of considering more important the set of the vertices that belong to the same class rather than the colors assigned to each vertex. Via a round robin criterion two bees are selected for generating one offspring: the biggest colorclass of the two selected parents is copied into the new solution, and its vertices are removed from the color classes of the belonging parent. This process is performed until classes with only one vertex are encountered. In this case, the single node is inserted inside one of the existing classes. In O-BEE-COL we have designed an optimized version of GPX, which differs from the original one basically in two aspects: (1) the number of solutions involved is determined by a parameter *partSol*; and (2) the cardinality of the colorclasses that must be copied into the new solution is determined by a parameter (*partLimit*). All colorclasses with cardinality greater or equal to *partLimit* will be copied inside the new solution. In this way, we want to force the transmission only of the best colorclasses to the offsprings. An experimental study conducted on the optimized GPX, also respect to the original one, confirmed us how these novelties introduced contribute significantly better on its performances (see plots in Fig. 3). The third novelty introduced in this work is the design of a Temperature mechanism that has the aim to dynamically self-handle some parameters during the evolution. The parameters bound to this self-regulating mechanism are: (1) number of parents involved in optimized GPX (*partSol*); (2) number of the improvement trails needed before to replace a solution (*evLimit*); (3) number of scout bees (*nScouts*); and (4) percentage of solutions that must be generated by randomized RLF during the scout bees phase (*percSol*). Whenever a better solution than the current one is found, the temperature mechanism sets the controlled parameters with their highest possible values, respectively $[100, 20, 5, 100\%]$. During the evolution, if no improvements occurred, then these values gradually decrease generation to generation until to reach their minimal values, which correspond to $[10, 5, 2, 10\%]$.

3   Results

In order to understand how the developed algorithm works, and how much is the contribution given by the novelties introduced we have performed many experiments using the classical DIMACS challenging benchmark[1]. O-BEE-COL has been tested on 22 instances (the most used), and it was compared with several algorithms, which represent the current state of the art for graph coloring problem. In this section we present all studies and experiments conducted, showing best tuning of the parameters; the impact factor contribution of the novelties designed; analysis on the running time; and comparisons conducted versus several algorithms. In most of the instances tested O-BEE-COL has found the best coloring, showing a robust convergence, and very competitive performances with respect the state of the art.

O-BEE-COL dynamics. One of the main goal when someone designs a generic EAs is to understand which is the best setting of the parameters because they strongly influence the performances of the algorithm. Thus many experiments have been performed with the aim to identify the best values of the parameters. As described in Sect. 2, O-BEE-COL depends on three parameters: population size ($popSize \in \{200, \ 500, 1000, 1500, 2000\}$); the lowest cardinality of the color classes allowed to be transmitted during the partitioning phase ($partLimit \in \{5, 10, 15, 18\}$); and the percentage of Employed Bees ($percEmp \in \{10\%, 20\%, 50\%, 70\%, 90\%\}$). To carefully analyse the proper tuning of the parameters, we

---

[1] http://mat.gsia.cmu.edu/COLOR/instances.html

conducted our study over several DIMACS instances, and for each combination of values we performed 10 independent runs. In Fig. 1 we show the convergence of O-BEE-COL on the instance $DSJC250.5$ since it is challenging enough to make robust our study. Inspecting all 100 experiments over this instance, O-BEE-COL obtains the best performances in term of success rate (SR) with the combination $(200, 5, 10\,\%)$. Due to a limit space, we show for each parameter the convergence plots produced in combination with the other two best values. Analysing the left plot (varying $popSize$) is possible to see how with large population size, O-BEE-COL quickly gets down towards low values within few generations, after which it shows a steady-state. On the other hand, choosing small dimensions, albeit the algorithm needs more generations, it achieves still the best coloring. However, inspecting step-by-step the convergence for each value, $popSize = 200$, although is the slowest, it is the one that performs a better exploration of the search space with the result of producing a good trade-off for diversity into the population. In the middle plot, are shown the convergence curves produced varying the parameter $partLimit$. The lower bound to the color classes transmitted during the partitioning phase is the one that contributes most to the convergence speed of the algorithm, and it usually assumes values within the range $\left(2, \frac{|V|}{\chi}\right)$. In particular, assigning $partLimit = 5$, O-BEE-COL has a slower convergence but it reaches the best solution before than the others. In the right plot, and last of Fig. 1, is shown the contribution given by $percEmp$, which indirectly represents the exploitation phase of the best solutions found so far. For all curves, O-BEE-COL shows a good trend without presenting fast or slow convergences. Comparing the curves between them is possible to see how O-BEE-COL with low percentage of employed bees is able to better explore the search space, and, at the same time, exploit better the information gained so far. In fact, with the lowest percentage possible ($percEmp = 10\,\%$) the algorithm achieves the best solution before than the others. It is important to point out how the best values for the three parameters correspond to their minimal values tested. This indicates us that there exists a good balance of diversity into the population, which helps the algorithm to get out from local optima.

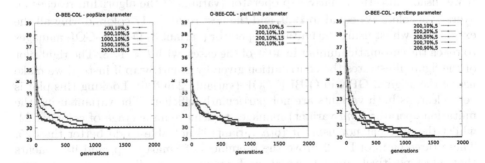

Fig. 1. Convergence behavior at varying the parameters: $popSize$, $partLimit$, and $percEmp$.

Table 1. Operating variants of O-BEE-COL, where $\hat{k}$ is the mean of the best colors found; k is the best coloring found in all runs; SR is the success rate, and AES is the average number of fitness function evaluations to the solution.

Variant	SmartSwap	Crossover	Temperature	$\hat{k}$	k	SR	AES
1	on	opt GPX	on	15	15	100 %	5, 972, 925
2	on	GPX	on	24	24	100 %	1, 503, 756
3	on	opt GPX	off	17.8	15	40 %	36, 599, 035
4	on	GPX	off	25	25	100 %	5
5	off	opt GPX	on	15.9	15	50 %	25, 981, 420
6	off	GPX	on	24	24	100 %	1, 639, 403
7	off	opt GPX	off	19.9	17	20 %	15, 872, 834
8	off	GPX	off	25	25	100 %	4

Several experiments have been conducted on the instance $le450\_15c$ in order to prove the effectiveness and utility of the features introduced in O-BEE-COL in terms of number of colors found; success rate; and average number of fitness function evaluations to the solution (AES). The aim of these experiments is to show that whatever the operators' combination chosen if we inhibit one of them, then its outcome will be negatively affected by this move. In Table 1 we show for any possible combination the average of the colors found $(\hat{k})$, best coloring found (k), SR and AES. In the next figures (Figs. 2, 3, and 4) we show a comparison of the several possible cases gradually disabling all the aforementioned features. The experiments have been averaged over 10 runs with different seeds. In the left plot of Fig. 2, we present the comparison of the convergence speed of O-BEE-COL with and without the SmartSwap operator (variants 1 and 5 of Table 1). It is possible to see how the first variant managed to reach the χ of the instance in every execution $(SR = 100\%)$, whilst turning off the SmartSwap operator, O-BEE-COL is able to get the best coloring only in 50 % of the executions. Middle plot shows a version of the algorithm that does not use the temperature mechanism. If we disable also the SmartSwap operator (variant 7) the algorithm reaches an average of colors $(\hat{k})$ equal to 19.9, and the best result of 17 colors during all the executions; whilst using the mutation operator (variant 3) O-BEE-COL manages to reach the chromatic number in 40 % of the cases, with $\hat{k} = 17.8$. The right plot of the figure illustrates the contribution given by SmartSwap if instead we make use of the original GPX in O-BEE-COL (variants 2 and 6). Looking this plot is very clear, as both variants are not particularly efficient. The variant using the mutation operator ($2nd$ variant) manages to achieve an average of colors of 24, whilst the one that not using it ($6th$ variant) is not able to do better than 25. These three plots of Fig. 2 prove the usefulness of SmartSwap, and its benefits that affect positively on the overall performances, regardless on the operators combination enabled. The plots in Fig. 3 prove the real goodness of the optimized GPX proposed with respect to the original version [15] improving significantly the performances of O-BEE-COL. The first plot on the left, presents a comparison of the speed convergences of O-BEE-COL using the proposed optimized crossover (1st variant) versus the original one (2nd variant). This comparison has been done on the fully enabled version of O-BEE-COL. The same comparison has been made also for the versions where the two other operators have been disabled (7th and 8th variants), and it is shown in the second plot on the left of the figure. Looking both plots becomes very clear as the developed optimized version to equality of variant outperforms significantly the original one. The last two plots in Fig. 3 show respectively the analysis conducted when we turn off the temperature mechanism (penultimate plot), and SmartSwap mutation operator (last plot). The role played by the optimized GPX is clearly evident even in these plots. In particular, disabling the Temperature mechanism or SmartSwap operator, O-BEE-COL with the original version of GPX is not able to achieve a coloring with less than 25 colors; whilst with the designed GPX version O-BEE-COL performs better decreasing the colors number in average to $\hat{k} = 17.8$ (with only temperature enabled) and $\hat{k} = 15.9$ (with only mutation operator enabled). Finally in Fig. 4 we show the improvements produced, in using the temperature mechanism, which controls dynamically the values of some parameters. In the left plot of Fig. 4 is plotted the difference concerning of O-BEE-COL with, and without the temperature mechanism. In both variants the algorithm achieves successfully the chromatic number, $\chi = 15$ (see Table 1). However, whilst the fully enabled version is able to achieved always the chromatic number (variant 1), when this operator is turned off (variant 3) the algorithm manages to achieve the best coloring only in 40 % of the executions. In middle plot the two different versions of the algorithm make no use of the mutation operator. When the temperature mechanism is enabled (5th variant) the algorithm finds the optimal coloring in one out of two cases ($\hat{k} = 15.9$), whilst the other combination (7th variant) does not manage to do better than a 17-coloring ($\hat{k} = 19.9$). The right plot shows the behavior of the algorithm using the classical version of GPX (2nd variant vs. 4th). Despite the poor performances, O-BEE-COL obtains a slightly better result when using the temperature mechanism (variant 2). In the overall, inspecting all combinations in Table 1 is possible to claim that the Temperature mechanism developed is the one that gives a positive greater contribution with respect to SmartSwap mutation operator.

Time-To-Target plots [1] have been used for studying the running time of O-BEE-COL, comparing the empirical and theoretical distributions. They represent a classical tool for characterizing the running time of stochastic algorithms in order to solve a specific optimization problem. In particular, we have used a Perl program proposed in [2], which display the probability that an algorithm will find a solution as good as a target within a given running time. Through this program two kinds of plots are produced: $QQ-$plot with superimposed variability information, and superimposed empirical and theoretical distributions. This kind of analysis has been conducted on the instances *School1* and *DSJC250.1*, performing 200 independent runs for each instance. The produced plots are shown in Fig. 5 (1st and 3rd plots for the first instance; 2nd and 4th plots for the last).

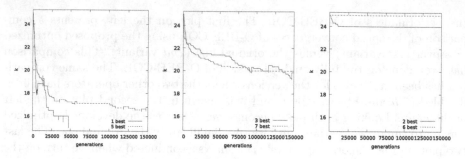

Fig. 2. Experimental analysis on the benefits provided by SmartSwap mutation operator.

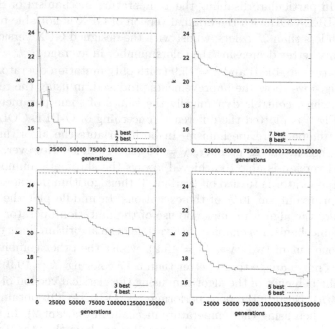

Fig. 3. Experimental analysis on the benefits provided by optimized GPX.

The plots show how for O-BEE-COL the empirical curve perfectly fits the theoretical one in both instances, except for very few worst cases (first two plots on the left). In the quantile-quantile plots, the O-BEE-COL results are in most of the cases equal to the theoretical ones, albeit a few less in *DSJC250.1* instance. This is explained because this last instance is more complex than the other one.

Experimental Comparisons. In order to evaluate the overall performances of O-BEE-COL, we have performed several experiments using the most known instances of the DIMACS benchmark [22]. The results in term of coloring found, *SR* obtained and *AES* needed are showed in Table 2. In this table we report for each instance its complexity characteristics; the chromatic number (χ); the

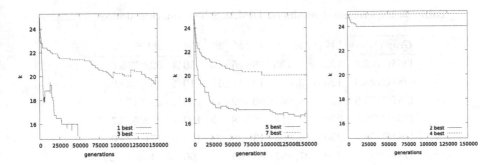

Fig. 4. Experimental analysis on the benefits provided by Temperature mechanism.

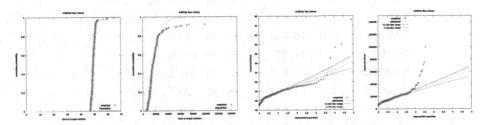

Fig. 5. Time to target plots for O-BEE-COL. The values have been obtained over 200 executions of the algorithm, respectively on the instance *School1* (*1st* and *3rd*) and *DSJC250.1* (*2nd* and *4th*).

best coloring known in literature (k^*); the best colors number found by O-BEE-COL (k), with *SR* and *AES* obtained. Each experiment has been performed on 10 independent runs. Inspecting such table, O-BEE-COL performs well on all instances *queen* and *school* finding the optimal coloring with a success rate of 100%. On the class of the instances *DSJC*, instead, O-BEE-COL seems to have more difficulty in getting the best coloring known, except for *DSJC125.1*, where it manages to find the optimal solution in only 5 tests out of 10, and for *DSJC125.5* where only in one case out of 10 the algorithm finds a 17-coloring. On the instances *DSJC250.1* and *DSJC205.5*, instead, the algorithm finds as best solution a coloring with only one color in more; whilst for the instances *DSJC125.9* and *DSJC250.9* the difference with the best coloring known is of 2 and 3 colors respectively. The same performances are achieved also in *le450_15* family, where O-BEE-COL achieves the chromatic number in *le450_15c* and *le450_15d* instances, whilst for the other two its solution differs from the chromatic number only for one color in more. Finally, in *flat300_20* and *flat300_26* O-BEE-COL founds the chromatic number producing a success rate of 100%, whilst in the last instance, *flat300_28*, it reaches a 31-coloring in 2 cases out of 10, where the chromatic number is however 28.

In Table 3 we present a comparison of O-BEE-COL with 6 different algorithms for the graph coloring problem, 4 of which nature-inspired: HPSO [30]; HCA [15]; GPB [20]; VNS [4]; VSS [21]; HANTCOL [13] (see the relative publications for major details). The best results are highlighted in boldface. Inspecting

Table 2. Experimental results on DIMACS benchmark instances [11, 22].

Graph	$|V|$	$|E|$	χ	k^*	k	SR	AES
DSJC125.1	125	736	5	5	5	50 %	528, 715.6
DSJC125.5	125	3, 891	12	17	17	10 %	464, 633.0
DSJC125.9	125	6, 961	30	42	44	100 %	29, 817.4
DSJC250.1	250	3, 218	8	8	9	100 %	252, 538.7
DSJC250.5	250	15, 668	13	28	29	100 %	471, 823.0
DSJC250.9	250	27, 897	35	69	73	90 %	24, 403, 325.4
le450_15a	450	8, 168	15	15	16	100 %	17, 678, 139.9
le450_15b	450	8, 169	15	15	16	100 %	6, 188, 035.6
le450_15c	450	16, 680	15	15	15	100 %	5, 972, 925.6
le450_15d	450	16, 750	15	15	15	80 %	18, 630, 401.3
flat300_20	300	21, 375	20	20	20	100 %	4, 800
flat300_26	300	21, 633	26	26	26	100 %	72.9K
flat300_28	300	21, 695	28	28	31	20 %	5.6M
Queen5_5	25	320	5	5	5	100 %	1.9
Queen6_6	36	580	7	7	7	100 %	1, 741.66
Queen7_7	49	952	7	7	7	100 %	6, 636.84
Queen8_8	64	1, 456	9	9	9	100 %	22, 107.25
Queen8_12	96	2, 736	12	12	12	100 %	1, 212, 000.35
Queen9_9	81	1, 056	10	10	10	100 %	31, 243.28
School1.nsh	352	14, 612	14	14	14	100 %	1, 703.28
School1	385	19, 095	14	14	14	100 %	821.5

this table is possible to see how the performances of O-BEE-COL are competitive with the compared algorithms, achieving in all tested instances the best coloring except in *DSJC250.5*. Moreover, albeit on *flat300_28* the VSS algorithm has found the lower number of colors, O-BEE-COL achieves yet the same results as all others.

In Table 4, O-BEE-COL is compared with other 10 algorithms: IMMALG [11, 28], MACOL [33], IGrAl [7], ACS [9], FCNS [29], IPM [14], ABAC [6], LAVCA, TPA and AMACOL [32]. The comparison has been performed with respect to the best coloring found. We have highlighted in boldface the colors found by O-BEE-COL, which are better or equal to the ones compared. Due a limit space, we refer the reader to each publication for more details on the algorithms. Also on these experiments is possible to see how O-BEE-COL is comparable with the state-of-the-art achieving the best coloring in 14 instances over 21. In the remaining instances nevertheless it isn't the worst.

Table 3. O-BEE-COL versus six different algorithms for graph coloring problem, with respect the best coloring found. The best results are highlighted in boldface.

Graph	O-BEE-COL	HPSO	HCA	GPB	VNS	VSS	HANTCOL
DSJC250.5	29	**28**	**28**	**28**	-	-	**28**
flat300_26	**26**	**26**	-	-	31	-	-
flat300_28	31	31	31	31	31	**29**	31
le450_15c	**15**	**15**	**15**	**15**	**15**	**15**	**15**
le450_15d	**15**	**15**	-	-	**15**	**15**	-

Table 4. O-BEE-COL versus state-of-the-art for graph coloring problem, with respect the best coloring found. The best or equal coloring obtained by O-BEE-COL is highlighted in boldface.

Graph	O-BEE-COL	IMMALG	MACOL	IGrAl	ACS	FCNS	IPM	ABAC	LAVCA	TPA	AMACOL
DSJC125.1	**5**	5	5	5	5	5	6	5	5	5	5
DSJC125.5	**17**	18	17	17	17	18	19	17	17	19	17
DSJC125.9	44	44	44	43	44	44	45	44	44	44	44
DSJC250.1	9	9	8	8	8	–	10	8	8	8	8
DSJC250.5	29	28	28	29	29	–	–	29	28	30	28
DSJC250.9	73	74	72	72	73	–	75	72	72	72	72
flat300_20_0	**20**	20	20	–	20	–	–	–	–	–	–
flat300_26_0	**26**	27	26	–	02	–	–	–	–	–	–
flat300_28_0	31	32	29	–	32	–	–	–	–	–	–
le450_15a	16	15	15	15	16	–	–	15	15	15	15
le450_15b	16	15	15	15	16	–	17	15	15	15	15
le450_15c	**15**	15	15	16	15	–	17	15	15	15	15
le450_15d	**15**	16	15	16	15	–	–	15	15	15	15
Queen5_5	**5**	5	–	5	–	–	–	5	–	–	–
Queen6_6	**7**	7	–	7	7	–	–	7	–	–	–
Queen7_7	**7**	7	–	7	7	–	–	7	–	–	–
Queen8_8	**9**	9	–	9	9	9	9	9	–	–	–
Queen8_12	**12**	12	–	12	12	–	–	12	–	–	–
Queen9_9	**10**	10	–	10	10	10	10	10	–	–	–
school1_nsh	**14**	15	14	14	14	–	–	14	–	–	–
School1	**14**	14	14	14	14	–	–	14	–	–	–

4   Conclusion

In this research paper we have developed a new Artificial Bee Colony heuristic, called O-BEE-COL, for the graph coloring problem. The novelties introduced in O-BEE-COL are basically: (1) SmartSwap mutation, which attempts to reduce the number of colorclasses, working primarily on the troublesome vertices; (2) optimized version of GPX, which works as multi-parents operator, forcing the transfer of the best colorclasses to the offsprings; and a (3) Temperature mechanism, which has the aim to dynamically handle some parameters.

Many experiments have been performed with the primary aim to evaluate the contribution, and benefits given by these new operators. Thus, all possible combinations of these three operators have been taken into account, and have been tested; the obtained results prove us how inhibiting one of them the overall performances are negatively affected. In particular, we show, via figures, the significant improvements produced by the optimized version of GPX, and as the Temperature mechanism is the one that gives a greater positive contribution, respect to the SmartSwap operator. Via Time-To-Target plots are also analysed the running times of O-BEE-COL, comparing the empirical and theoretical curves. Finally, a comparison with the state-of-the-art has been conducted as well, in order to evaluate the robustness and efficiency of O-BEE-COL. Inspecting all results, and comparisons O-BEE-COL shows efficiency; robustness; and very competitive performances, achieving in the most of the instances the chromatic number, or the best coloring known.

References

1. Aiex, R.M., Resende, M.G.C., Ribeiro, C.C.: Probability distribution of solution time in GRASP: an experimental investigation. J. heuristics **8**, 343–373 (2002)
2. Aiex, R.M., Resende, M.G.C., Ribeiro, C.C.: TTTPLOTS: a perl program to create time-to-target plots. Optim. Lett. **1**, 355–366 (2007)
3. Akay, B., Karaboga, D.: A modified artificial bee colony algorithm for real-parameter optimization. Inf. Sci. **192**, 120–142 (2012)
4. Avanthay, C., Hertz, A., Zufferey, N.: A variable neighborhood search for graph coloring. Eur. J. Oper. Res. **151**(2), 379–388 (2003)
5. Bouziri, H., Mellouli, K., Talbi, E.-G.: The k-coloring fitness landscape. J. Comb. Optim. **21**(3), 306–329 (2011)
6. Bui, T.N., Nguyen, T.-V.H., Patel, C.M., Phan, K.-A.T.: An ant-based algorithm for coloring graphs. Discrete Appl. Math. **156**, 190–200 (2008)
7. Caramia, M., Dell'Olmo, P.: Coloring graphs by iterated local search traversing feasible and infeasible solutions. Discrete Appl. Math. **156**, 201–217 (2008)
8. Chow, F.C., Hennessy, J.L.: The priority-based coloring approach to register allocation. ACM Trans. Program. Lang. Syst. **12**, 501–536 (1990)
9. Consoli, P., Collerá, A., Pavone, M.: Swarm intelligence heuristics for graph coloring problem. In: IEEE Congress on Evolutionary Computation (CEC), vol. 1, pp. 1909–1916 (2013)
10. Costa, D., Hertz, A.: Ants can colour graphs. J. Oper. Res. Soc. **48**, 295–305 (1997)
11. Cutello, V., Nicosia, G., Pavone, M.: An immune algorithm with stochastic aging and kullback entropy for the chromatic number problem. J. Comb. Optim. **14**(1), 9–33 (2007)
12. de Werra, D.: An introduction to timetabling. Eur. J. Oper. Res. **19**, 151–162 (1985)
13. Dowsland, K.A., Thompson, J.M.: An improved ant colony optimisation heuristic for graph colouring. Discrete Appl. Math. **156**(3), 313–324 (2008)
14. Dukanovic, I., Rendl, F.: A semidefinite programming-based heuristic for graph coloring. Discrete Appl. Math. **156**, 180–189 (2008)
15. Galinier, P., Hao, J.: Hybrid evolutionary algorithms for graph coloring. J. Comb. Optim. **3**(4), 379–397 (1999)
16. Gamst, A.: Some lower bounds for a class of frequency assignment problems. IEEE Trans. Veh. Technol. **35**, 8–14 (1986)
17. Garey, M.R., Johnson, D.S.: Computers and Intractability: a Guide to the Theory of NP-completeness. Freeman, New York (1979)

18. Garey, M.R., Johnson, D.S., So, H.C.: An application of graph coloring to printed circuit testing. IEEE Trans. Circuits Syst. **CAS–23**, 591–599 (1976)
19. Glass, C.: Bag rationalization for a food manufacturer. J. Oper. Res. Soc. **53**, 544–551 (2002)
20. Glass, C.A., Prügel-Bennet, A.: Genetic algorithm for graph coloring: exploration of Galinier and hao's algorithm. J. Comb. Optim. **7**(3), 229–236 (2003)
21. Hertz, A., Plumettaz, M., Zufferey, N.: Variable space search for graph coloring. Discrete Appl. Math. **156**, 2551–2560 (2008)
22. Johnson, D.S., Trick, M.A.: Cliques, Coloring and Satisfiability: Second DIMACS Implementation Challenge. American Mathematical Society, Providence (1996)
23. Karaboga, D., Basturk, B.: A powerful and efficient algorithm for numerical function optimization: artificial bee colony (ABC) algorithm. J. Global Optim. **39**(3), 459–471 (2007)
24. Karaboga, D., Basturk, B.: On the performance of Artificial Bee Colony (ABC) algorithm. Appl. Soft Comput. **8**, 687–697 (2008)
25. Krasnogor, N., Smith, J.E.: A tutorial for competent memetic algorithms: model, taxonomy and design issues. IEEE Trans. Evol. Comput. **9**(5), 474–488 (2005)
26. Leighton, F.T.: A graph coloring algorithm for large scheduling problems. J. Res. Natl. Bur. Stan. **84**, 489–505 (1979)
27. Oner, A., Ozcan, S., Dengi, D.: Optimization of university course scheduling problem with a hybrid artificial bee colony algorithm. In: IEEE Congress on Evolutionary Computation, pp. 339–346 (2011)
28. Pavone, M., Narzisi, G., Nicosia, G.: Clonal selection - an immunological algorithm for global optimization over continuous spaces. J. Global Optim. **53**(4), 769–808 (2012)
29. Prestwich, S.: Generalised graph colouring by a hybrid of local search and constraint programming. Discrete Appl. Math. **156**, 148–158 (2008)
30. Qin, J., Yin, Y., Ban, X.-J.: Hybrid discrete particle swarm algorithm for graph coloring problem. J. Comput. **6**(6), 1175–1182 (2011)
31. Rodriguez, F.J., García-Martínez, C., Blum, C., Lozano, M.: An artificial bee colony algorithm for the unrelated parallel machines scheduling problem. In: Coello, C.A.C., Cutello, V., Deb, K., Forrest, S., Nicosia, G., Pavone, M. (eds.) PPSN 2012, Part II. LNCS, vol. 7492, pp. 143–152. Springer, Heidelberg (2012)
32. Torkestani, J.A., Meybodi, M.R.: A new vertex coloring algorithm based on variable action-set learning automata. Comput. Inform. **29**(1), 447–466 (2010)
33. Zhipeng, L., Hao, J.-K.: A memetic algorithm for graph coloring. Eur. J. Oper. Res. **203**(1), 241–250 (2010)

PSO with Tikhonov Regularization
for the Inverse Problem in Electrocardiography

Alejandro Lopez[1]([⊠]), Miguel Cienfuegos[2], Bedreddine Ainseba[1],
and Mostafa Bendahmane[1]

[1] Institut de Mathematiques de Bordeaux,
Universite Victor Segalen Bordeaux 2, Bordeaux, France
alejandro.lopezrn@hotmail.com
[2] Departamento de Electrónica, Universidad de Guadalajara,
CUCEI, Av. Revolución 1500, Guadalajara, Jal, Mexico

Abstract. In this paper we present a nearest neighbor particle swarm
optimization (PSO) algorithm applied to the numerical analysis of the
inverse problem in electrocardiography. A two-step algorithm is proposed based on the application of the modified PSO algorithm with the
Tikhonov regularization method to calculate the potential distribution
in the heart. The PSO improvements include the use of the neighborhood
particles as a strategy to balance exploration and exploitation in order
to prevent premature convergences and produce a better local search.
In the literature the inverse problem in electrocardiography is solved
using the minimum energy norm in a Tikhonov regularization scheme.
Although this approach solves the system, the solution may not have a
meaning in the physical sense. Comparing to the classical reconstruction,
the two-step PSO algorithm improves the accuracy of the solution with
respect to the original distribution. Finally, to validate our results, we
create a distribution over the heart by using a model of electrical activity
(Bidomain model) coupled with a volume conductor model for the torso.
Then, using our method, we make the reconstruction of the potential
distribution.

Keywords: Direct and inverse problems · Particle swarm optimization
(PSO) · Bio-inspired algorithm · Finite element · Electrocardiography ·
Bidomain

1 Introduction

Cardiovascular disease is the leading cause of mortality in the Western countries
and the most common cause of death in people beyond 35 years in China, India
and South America [1]. Although cardiac function is linked to its muscular contraction in the common minds, this mechanical function is fully determined and
dependent on prior electrical activation of the cardiac cells. Therefore any cardiac electrical disorder would impact on muscular contraction. The ideal solution

© Springer International Publishing Switzerland 2014
P. Legrand et al. (Eds.): EA 2013, LNCS 8752, pp. 256–270, 2014.
DOI: 10.1007/978-3-319-11683-9_20

will be to measure directly the potential in the heart, but this is highly invasive. To calculate the electrical activity on the heart using boundary surface potential measurements (BSPMs) is known as the inverse problem in electrocardiography.

The methodology is to consider the torso as a volume conductor (ruled by the equation of Laplace), and then using a high density electrocardiogram to measure the electrical activity on the thorax' surface. This problem is considered as an ill-posed boundary value, and it is commonly solved employing regularization techniques [2]. In the literature the heart is considered as a closed surface in a quasi-static scheme [3]. This approach has a linear relationship to the BSPMs [4], but it is not possible to determine the sources in the cardiac volume. Considering that the inverse problem in electrocardiography is ill-posed, many techniques and methods have been developed to constrain the possible solutions; stochastic search algorithms like genetic algorithm (GA) and particle swarm optimization (PSO) have been found to be effective in dealing with these type of problems [5,6].

PSO is a population-based evolutionary technique inspired by the social analogy of swarm behavior in populations of natural organisms, such as a flock of birds or a school of fish [7]. The main procedure for PSO is to generate a population of candidate solutions, called particles. The particles are moved in the search-space according to the mathematical formula of the particle's position and velocity. Each particle's movement is influenced by it's local best known position and toward the best known positions in the search-space. The positions are updated as better positions are found by other particles. This moves the swarm toward the best solutions. The final solution is chosen by a stop criterion or a specified number of iterations [8]. The PSO has been successfully applied in a wide variety of optimization and inverse problems [9], for example inverse scattering problems [10], for geophysical inverse problems [11], for inverse heat conduction problems [12]. Moreover in [13], the authors used the PSO to determine parameters to a predator-prey model.

The idea of optimization algorithms for the inverse problem in electrocardiography can be found in the literature. For example in [30], the authors propose the use of real-valued genetic algorithms for the estimation of multiple regularization parameters, that otherwise can not be measured. These parameters are used to constrain the solution spatially and temporally. In the work by [14] evolutionary algorithms are employed with a set of real measures, and regularized solutions to improve the solution of the inverse problem. This method is similar to the one proposed by [15] using artificial neural networks instead of an evolutionary approach. Both of these methods require training data sets for the algorithm to work. A similar approach can be found in [16], where the construction of the initial populations comes from Tikhonov regularized solutions.

Although PSO has proved to converge quickly towards an efficient solution in a reduced number of iterations [8], it has been reported that PSO experiences difficulties in reaching the global optimal solution in some optimization problems [17], and can suffer premature convergence [18]. In this paper, we use an improved PSO algorithm, using the nearest neighbor based on [19] to improve the performance of the standard PSO. Kennedy empirically examined the effects of some neighborhood topologies in the PSO [20]. Similar approaches have been proposed in [21] and in [22] showing better results compared to standard PSO.

The modified PSO was implemented in a two-step scheme to solve the inverse problem in electrocardiography, and reconstruct the cardiac sources. We generate an operator, which gives the relationship between the membrane potential on the heart, and the potential on the thorax surface using Finite Element Method approximation. The solution is in a two step algorithm. The first step is to utilize the enhanced PSO to create an approximate answer. The algorithm will look for the coefficients of the fundamental solution of the Laplace equation that solves the volume conductor system. In the second step, we will use the solution from the modified PSO in a Tikhonov regularization scheme [23], as a priori information. The system was tested by using voltage distributions generated by the Bidomain model (see for e.g. [24]). The Bidomain model is used to calculate forward computations of extra-cellular and BSPMs using membrane potentials in the heart. A set of membrane potentials are created, and then voltage distributions over the thorax are calculated using them. In our numerical tests the membrane potentials will be reconstructed and compared to the originals.

The paper is organized as follows. In Sect. 2, we explain some information on inverse problem in electrocardiography. Then Sect. 3 describes the particle swarm optimization (PSO) and the nearest neighbor PSO algorithm. In Sect. 4, we demonstrate the simulation experiments, and we discuss the results. Lastly, we present the conclusions.

2   Methods

2.1   Create the Operator

The Bidomain model is a model of the electrical properties of the cardiac muscle averaged over many cells. The model considers the anisotropy of the intracellular and extracellular domains, which affects the electrical behavior. The model is highly anisotropic; there will be different conductivities for the direction parallel, perpendicular and normal to the fiber directions of the cardiac muscle. The Bidomain model is given by the following equation in terms of u_e, and v_m, which are the extracellular and membrane potential:

$$\nabla \cdot (M_i \nabla (v_m + u_e)) = \chi C_m \frac{\partial v_m}{\partial t} + \chi I_{ion} + \chi I_{app}, \tag{1}$$

$$-\nabla \cdot (M_i \nabla v_m) = \nabla \cdot ((M_i + M_e) \nabla u_e). \tag{2}$$

where C_m is the membrane capacitance per unit area, χ is the membrane surface-to-volume ratio. The conductivity tensors for the intracellular, and extracellular medium are M_i, and M_e. The ionic current is given by I_{ion}, and the applied current is I_{app}. If we consider equal anisotropy rates $M_e = \lambda M_i$ [24] then we can reduce the system, to the simplified model (Monodomain), for a further explanation refer to [24]. The Monodomain model is given by the following set of equations:

$$\frac{\lambda}{1+\lambda} \nabla \cdot (M_i \nabla v_m) = \chi C_m \frac{\partial v}{\partial t} + \chi I_{ion} + \chi I_{app}, \tag{3}$$

$$\nabla \cdot (M_i \nabla v_m) = -\nabla \cdot ((1+\lambda) M_i \nabla u_e). \tag{4}$$

Observe that the Eq. (4) can be expressed numerically in the following matrix form:

$$M v_m = N u_e, \tag{5}$$

or

$$M N^{-1} v_m = u_e. \tag{6}$$

If we take the Laplace equation system, that describes the volume conductor model;

$$\begin{aligned} -\nabla \cdot (\kappa \nabla u) &= 0 \ \ in \ \Omega, \\ u &= g \ \ on \ \Gamma_1, \\ \kappa \nabla u n &= h \ \ on \ \Gamma_2, \end{aligned} \tag{7}$$

Herein, Ω is the torso, Γ_1 the heart surface, and Γ_2 the thorax surface. The nodes in the thorax will be indicated with sub-index t, the nodes in the heart h, and the nodes in between v. We can build the matrix-vector system in the following form. First we calculate the Stiffness Matrix; we calculate the stiffness matrix which entries are equal to

$$K_{ij} = \int_\Omega \kappa \nabla \phi_i \nabla \phi_j d\Omega \quad i, j = 1, 2, ..., N (number \ of \ nodes). \tag{8}$$

The resulting matrix vector equation will be:

$$\begin{bmatrix} K_{hh} & K_{hv} & K_{ht} \\ K_{vh} & K_{vv} & K_{vt} \\ K_{th} & K_{tv} & K_{tt} \end{bmatrix} \begin{bmatrix} u_h \\ u_v \\ u_t \end{bmatrix} = \begin{bmatrix} 0 \\ 0 \\ 0 \end{bmatrix}. \tag{9}$$

The nodal values of the potential are u_h, u_v, u_t for the inner surface, volume, and outer surface, respectively. Then we apply the Dirichlet condition; we consider u_h is given, and considering no overlapping between the surfaces. Then, the system becomes

$$\begin{bmatrix} K_{vh} & K_{vv} & K_{vt} \\ 0 & K_{tv} & K_{tt} \end{bmatrix} \begin{bmatrix} u_h \\ u_v \\ u_t \end{bmatrix} = \begin{bmatrix} 0 \\ 0 \end{bmatrix}, \tag{10}$$

or

$$\begin{bmatrix} K_{vv} & K_{vt} \\ K_{tv} & K_{tt} \end{bmatrix} \begin{bmatrix} u_v \\ u_t \end{bmatrix} = \begin{bmatrix} -K_{vh} u_h \\ 0 \end{bmatrix}. \tag{11}$$

Next, we apply the Neumann condition. For each triangle in the outer surface (where the Neumann condition is applied), we calculate the following coefficient;

$$Neumann_c = (A)/3.0, \tag{12}$$

Where A is the area of the triangle. Then, we apply these contributions to each of the nodes of the triangle. In the end we will have a vector the size of the nodes of the outer surface. This vector we will call it N_v. The contributions will be in the form $N_{v_i} + = Neumann_{c_i}$. Creating this vector the global Laplace matrix-equation will be;

$$\begin{bmatrix} K_{vv} & K_{vt} \\ K_{tv} & K_{tt} \end{bmatrix} \begin{bmatrix} u_v \\ u_t \end{bmatrix} = \begin{bmatrix} -K_{vh}u_h \\ N_v^t \kappa \nabla u_t \cdot n \end{bmatrix}. \tag{13}$$

If the values in the volume in between are not from our interest; from the Eq. (13) we can build a direct relationship between the potentials on the two surfaces;

$$K_{vv}u_v + K_{vt}u_t = -K_{vh}u_h, \tag{14}$$
$$K_{tv}u_v + K_{tt}u_t = N_v^t \kappa \nabla u_t \cdot n. \tag{15}$$

Observe that from (14) we get

$$u_v = -K_{vv}^{-1}(K_{vh}u_h + K_{vt}u_t). \tag{16}$$

Using this in (15), we obtain,

$$- K_{tv}K_{vv}^{-1}K_{vh}u_h - K_{tv}K_{vv}^{-1}K_{vt}u_t + K_{tt}u_t = N_v^t \kappa \nabla u_t \cdot n, \tag{17}$$

or

$$(K_{tt} - K_{tv}K_{vv}^{-1}K_{vt})u_t = N_v^t \kappa \nabla u_t \cdot n + K_{tv}K_{vv}^{-1}K_{vh}u_h. \tag{18}$$

We will define the operators P and Q as follows:

$$P = (K_{tt} - K_{tv}K_{vv}^{-1}K_{vt})^{-1}K_{tv}K_{vv}^{-1}K_{vh}, \tag{19}$$
$$Q = (K_{tt} - K_{tv}K_{vv}^{-1}K_{vt})^{-1}N_v^t. \tag{20}$$

Then we can write the system (18) in the following form

$$u_t = Pu_h + Q\kappa \nabla u_t \cdot n. \tag{21}$$

The flux over the thorax is null so the multiplication $Q\kappa \nabla u_t \cdot n$, will be zero. Thus, the relationship between the outer and inner surface will be

$$Pu_h = u_t. \tag{22}$$

For our test we consider an isolated heart; this means there is no continuity of the flux from the heart, and we take into account the potential as a Dirichlet condition. For this we add the following relationship

$$u_h = u_e. \tag{23}$$

The result from this and (6) is

$$PMN^{-1}v_m = u_t. \tag{24}$$

Since the inverse problem is an ill-posed problem a regularization technique is necessary. The regularization technique used in our study is a global Tikhonov-scheme. For this global scheme, the nodal values u_h can be estimated by minimizing a generalized form of the discretized Tikhonov functional:

$$min_{v_m}(||PMN^{-1}v_m - u_t||^2 + \lambda||C(v_m - v'_m)||^2), \ \mu > 0, \qquad (25)$$

where C is a constrained matrix (the identity matrix), and v'_m is the a priori information. For the minimum energy $v'_m = 0$, and for the two-step $v'_m = v_{pso}$; where v_{pso} is an approximation made using particle swarm optimization. For our datasets, we took the value of $\mu = 0.00001$.

3   Two-Step Algorithm

3.1   Particle Swarm Optimization

Particle Swarm Optimization (PSO) is a population-based evolutionary algorithm based on the social behaviour of birds flocking and fish schooling, it was firstly introduced by [7]. The population denominated as swarm uses a number of particles (candidate solutions) which are moved around the search space to find best solution using their positions. Each particle cooperates with the others during the search process by sharing the information of its current position with the best position that it and the other particles in the swarm have found. The mathematical formulation of the PSO is as follows:

Initially, a number of particles N of the swarm x_i are randomly positioned in the search space and random velocities v_i are assigned to each particle. Then, each particle is evaluated by calculating the objective function. Once the particles have been evaluated the values of the particle's best position p_i and the global best position g are calculated. Next, the algorithm iterates until the stopping criterion is met; that is either an acceptable minimum error is attained or the maximum number of iterations is exceeded. In each k iteration, each particles position x_i^{k+1} and velocity v_i^{k+1} are updated following the next equations:

$$v_i^{k+1} = \omega \cdot v_i^k + c_1 \cdot r_1 \left(p_i^k - x_i^k\right) + c_2 \cdot r_2 \left(g^k - x_i^k\right) \qquad (26)$$

$$x_i^{k+1} = x_i^k + v_i^{k+1}, \qquad (27)$$

where ω is a real constant called inertia weight, c_1 and c_2 are the acceleration coefficients that moves the particles toward the local and global best positions; and r_1 and r_2 are both random values uniformly distributed between zero and one. The process is repeated until the stopping condition is met, the final value of g^k represents the optimum solution found for the problem optimized using this algorithm.

3.2   Nearest Neighbor Particle Swarm Optimization

As mentioned above, an enhanced PSO is used based in local neighborhood topology. A third term in the velocity calculation is aggregated using the nearest neighbor in the search space. The nearest neighbor rule is based on the distance between particles in the search space. The result equation by adding the nearest neighbor n_i^k to the calculation of velocity in each particle is:

$$v_i^{k+1} = \omega \cdot v_i^k + c_1 \cdot r_1 \left(p_i^k - x_i^k\right) + c_2 \cdot r_2 \left(g^k - x_i^k\right) + c_3 \cdot r_3 \left(x_i^k - n_i^k\right). \quad (28)$$

The nearest neighbor n_i^k is the nearest individual according to a distance computed on the swarm, n_i^k is modeled according to the following equation:

$$n_i^k = \min_{j \in \{1,2,\ldots,N\}} (d_{i,j}), \quad (29)$$

where the $d_{i,j}$ is the Euclidean distance between the particles i and j, such that $d_{i,j} = \|x_i - x_j\|$.

The ratio between p_i^k, g^k and n_i^k controls the effect of the velocities and the trade-off between the global and local exploration capabilities of PSO. The additional term can be considered as a vibration. At first steps in the algorithm, the distance between two arbitrary particles is large, and the vibration may produce a better probabilities to escape from local minimums. At final stages, the distance of two arbitrary particles is small, this may provide a local exploitation in the area. These aspects, avoiding premature convergence and local search can make the nearest neighbor PSO algorithm converge more efficiently to global optimum.

3.3   Two-Step Algorithm Using Nearest Neighbor PSO

The optimization process begins by setting a random set of possible solutions with a fixed initial number of members in the swarm, called particles. In the swarm each particle is defined by a collection of variables. The solution will have the form of the fundamental solution of Laplace. The fundamental solution of the Laplace equation in 3D centered at a point (ξ, η, ς) is:

$$\frac{\partial^2 \omega}{\partial x^2} + \frac{\partial^2 \omega}{\partial y^2} + \frac{\partial^2 \omega}{\partial z^2} + \delta(x - \xi, y - \eta, z - \varsigma) = 0. \quad (30)$$

or

$$\omega(x) = -\frac{A}{4\pi \sqrt{(x - \xi)^2 + (y - \eta)^2 + (z - \varsigma)^2}} + B, \quad (31)$$

where ξ, η, ς, A and B are constants. Note that from (31) we have 5 coefficients to find (ξ, η, ς, A and B).

The parameters of both PSO and the modified PSO are set to $c_1 = 2$, $c_2 = 2$; besides, the weight factor decreases linearly from 0.9 to 0.2 [8]. The other parameters has been determined experimentally, they are kept for all experiments. Such parameters are set to $N = 50$, $c_3 = 0.5$ and iteration number $= 1000$. The steps involved in the nearest neighbor PSO algorithm are detailed below. Initially the maximum (max) and minimum (min) limits in the search space are defined for each value. For the membrane potential v_m will be -85 mv, and 15 mv respectively. Each particle of the swarm will be a vector containing the

5 values of the coefficients. The particles are composed of 5 decision variables. Each particle is evaluated using (31) at each node, creating solution vectors. The details steps for the nearest neighbor PSO are listed as follows.

Step 1. The algorithm randomly initialize positions and velocities for all of the particles in the swarm. Include one particle initialized at -85, which is the stability value for the electrical activity.

Step 2. The vectors are evaluated by the norm

$$||PMN^{-1}v_m - u_t||^2. \tag{32}$$

Step 3. The personal historical best position p_i^k, the global best position g^k and every nearest neighbor position n_i^k are updated.

Step 4. At iteration k, the velocity of the particle i, is updated as:

$$v_i^{k+1} = \omega \cdot v_i^k + c_1 \cdot r_1 \left(p_i^k - x_i^k\right) + c_2 \cdot r_2 \left(g^k - x_i^k\right) + c_3 \cdot r_3 \left(x_i^k - n_i^k\right) \tag{33}$$

and the new position is computed as:

$$x_i = x_i + v_i \tag{34}$$

Step 5. The procedure is repeated until

$$||PMN^{-1}v_m - u_t||^2 < \beta \text{ or } n_{iter} < 1000, \tag{35}$$

where β is the stop parameter value, and n_{iter} is the number of iterations.

Step 6. Use the nearest neighbor PSO result as a priori information in Eq. (25) $v_m' = v_g$.

The overall method is the following:

- *Create transfer matrix using FEM.*
- *Create operator for the relationship between membrane potential and BSPMs.*
  - *$for(i = 0; i < Measures\ quantity; i++)$*
    - *Create 99 first vector solutions randomly.*
    - *Create 1 vector solution at -85.*
    - *$while(||PMN^{-1}v_m - u_t||^2 < \beta \text{ or } n_{iter} < 1000)$*
      - *Evaluate particles.*
      - *Update historical best, global best and nearest neighbor positions of every particle.*
      - *Calculate velocities.*
      - *Update positions.*
    - *$min_{v_m}(||PMN^{-1}v_m - u_t||^2 + \lambda||C(v_m - v_m')||^2), v_m = v_g.$*

4 Experimentation

In order to validate the two-step algorithm, we generate a voltage distribution in the thorax by using the Bidomain model (to resemble the real electrical activity of the heart). Later, we rebuild the membrane potential using the minimum energy norm $v'_m = 0$, and the nearest neighbor PSO solution $v'_m = v_g$. To assess the precision of the reconstructed solution we use the following formula

$$difference = \frac{\sum(v_{m_i} - v_{m_i}*)^2}{\sum(v_{m_i})^2}, \tag{36}$$

for the difference between the original distribution v_m and the calculated one v_m*.

4.1   Experiment 1

In the first test an impulse in the basal plane over the left ventricle was applied. The original membrane potential generated with the cardiac model is in Fig. 1. The reconstructed model using the minimum energy norm is found in Fig. 2. The resulted model from the two-step algorithm is shown in Fig. 3. The heart is inverted showing the basal plane in the bottom, and the apex on the top for visualization purposes.

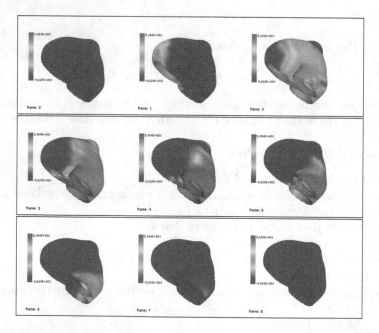

Fig. 1. Original Membrane Potential Distribution for 0 ms, 50 ms, 150 ms, 200, ms, 250 ms, 300 ms, 350 ms, 400 ms created using the Bidomain model for one pulse.

4.2   Experiment 2

In the second experiment impulses in three points over the basal plane were applied; in the left ventricle, in the right ventricle and the wall that divides them.

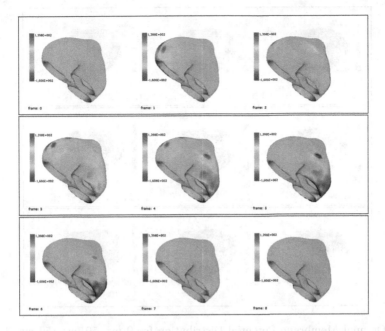

Fig. 2. Membrane Potential Distribution for 0 ms, 50 ms, 150 ms, 200, ms, 250 ms, 300 ms, 350 ms, 400 ms originated using the minimum energy norm.

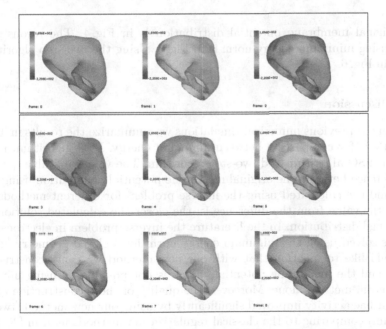

Fig. 3. Membrane Potential Distribution for 0 ms, 50 ms, 150 ms, 200, ms, 250 ms, 300 ms, 350 ms, 400 ms originated using the two-step algorithm.

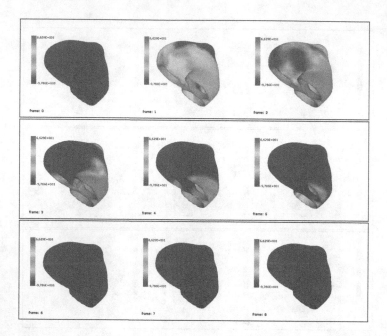

Fig. 4. Original Membrane Potential Distribution for 0 ms, 50 ms, 150 ms, 200, ms, 250 ms, 300 ms, 350 ms, 400 ms created using the Bidomain model for three pulses over the basal plane.

The original membrane potential distribution is in Fig. 4. The reconstructed model using minimum energy norm is in Fig. 5. Using the two-step algorithm is shown in Fig. 6.

4.3   Discussion

Based in the previous numerical simulations we summarize the results in Table 1 (using (36)) for comparison between minimum energy, PSO algorithm, nearest neighbor PSO algorithm and two-step algorithm. The values in Table 1, refer to the difference between the original membrane potential distribution using Bidomain, and the calculated using the inverse problem for different methods. The smaller the value from the difference is; the closer the calculated solution is to the original distribution. In the literature the inverse problem in electrocardiography is solved using the minimum energy norm for a closed geometry [25–28]. We would like to mention that with our new method we can reconstruct the sources and the membrane potential, which is not possible by using minimum energy norm method alone. Moreover, the quality of the reconstruction of electrical cardiac activity improved significantly by using our new method (two-step algorithm) comparing to the classical regularization methods used in [26–28].

Table 1 summarizes the results obtained, all algorithms have been programmed in C# over the same computer. The simulations have been executed 10 times

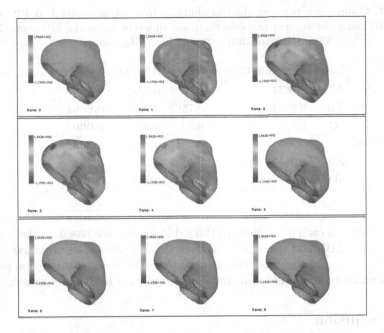

Fig. 5. Membrane Potential Distribution for 0 ms, 50 ms, 150 ms, 200, ms, 250 ms, 300 ms, 350 ms, 400 ms created using the minimum energy norm.

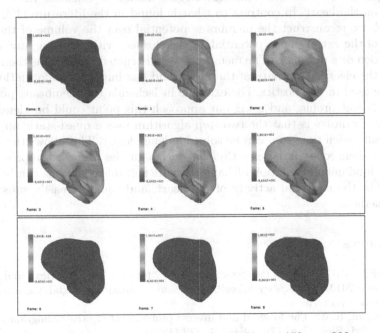

Fig. 6. Membrane Potential Distribution for 0 ms, 50 ms, 150 ms, 200, ms, 250 ms, 300 ms, 350 ms, 400 ms created using the two-step algorithm.

Table 1. Comparison between the distributions from experiments 1 and 2, and the calculated using the inverse problem for these different methods; Minimum Energy, PSO Algorithm, NN-PSO Algorithm and Two-Step Algorithm.

	Index	Minimum Energy	PSO	NN-PSO	Two-Step Algorithm
1 Pulse	A	0.9674	0.7341	0.6937	**0.6656**
	AB	0.9763	0.7562	0.7056	**0.6661**
	SD	0.11	0.24	0.005	**0.0004**
3 Pulses	A	0.9770	0.4341	0.3861	**0.3632**
	AB	0.9847	0.4359	0.3872	**0.3635**
	SD	0.13	0.34	0.003	**0.0002**

independently on each algorithm. In this table, results are based on the best (B), average best (AB), and standard deviation (SD) of the values obtained by each algorithm. As can be seen from results of Table 1, the two-step algorithm presents the best performance and obtains the best precision in both experiments.

5   Conclusion

In our paper, we proposed a novel two-step scheme algorithm using the nearest neighbor PSO with the Tikhonov Regularization to calculate the electrical sources on the heart. In contrast to what is found in the literature [29], in our approach we reconstruct the membrane potential over the volume of the heart instead of the extracellular potential on the surface, without using any a priori information or a database. The membrane reconstructed potential is the responsible of the electrical activity of the heart, and has important information that could be used in diagnostics. For example in Ischemia the membrane potential has a different profile, and using our approach this point could be identifiable. Another advantage is that the two step algorithm uses a quasi-static approach, numerically, each time step can be solved independently. This allows the system to be parallelized. It is noteworthy that although the proposed approach was used for fundamental solution of Laplace, more suitable equations can be found to describe the electrical activity of the heart, and could be easily substituted in the method.

References

1. IHME: GBD Cause Patterns. Institute for Health Metrics and Evaluation (2013). http://www.healthmetricsandevaluation.org/gbd/visualizations/gbd-cause-patterns
2. Gulrajani, R.M.: The forward and inverse problems of electrocardiography. IEEE Eng. Med. Biol. Mag. **17**(5), 84–101, 122 (1998)
3. Rudy, Y.: Noninvasive imaging of cardiac electrophysiology and arrhythmia. Ann. N. Y. Acad. Sci. **1188**, 214–221 (2010)

4. Yamashita, Y.: Theoretical studies on the inverse problem in electrocardiography and the uniqueness of the solution. IEEE Trans. Biomed. Eng. **BME–29**(11), 719–725 (1982)
5. Jiang, M., Huang, W., Xia, L., Shou, G.: The use of genetic algorithms for optimizing the regularized solutions of the Ill-posed problems. In: Proceedings of the 2008 Second International Symposium on Intelligent Information Technology Application, vol. 3, pp. 119–123. IEEE Computer Society (2008)
6. Chen, M.-Y., Hu, G., He, W., Yang, Y.-L., Zhai, J.-Q.: A reconstruction method for electrical impedance tomography using particle swarm optimization. In: Li, K., Fei, M., Jia, L., Irwin, G.W. (eds.) LSMS/ICSEE 2010. LNCS, vol. 6329, pp. 342–350. Springer, Heidelberg (2010)
7. Kennedy, J., Eberhart, R.: Particle swarm optimization. In: Proceedings of the IEEE International Conference on Neural Networks, vol. 4, pp. 1942–1948 (1995)
8. Shi, Y., Eberhart, R.C.: Empirical study of particle swarm optimization. In: Proceedings of the 1999 Congress on Evolutionary Computation, CEC 99, vol. 3 (1999)
9. Poli, R.: Analysis of the publications on the applications of particle swarm optimisation. J. Artif. Evol. Appl. (2008)
10. Donelli, M., Franceschini, G., Martini, A., Massa, A.: An integrated multiscaling strategy based on a particle swarm algorithm for inverse scattering problems. IEEE Trans. Geosci. Remote Sens. **44**(2), 298–312 (2006)
11. Fernández Martínez, J.L., García Gonzalo, E., Fernández Álvarez, J.P., Kuzma, H.A., Menéndez Pérez, C.O.: PSO: a powerful algorithm to solve geophysical inverse problems: application to a 1D-DC resistivity case. J. Appl. Geophys. **71**(1), 13–25 (2010)
12. Liu, F.-B.: Particle Swarm Optimization-based algorithms for solving inverse heat conduction problems of estimating surface heat flux. Int. J. Heat Mass Transf. **55**(78), 2062–2068 (2012)
13. Martínez-Molina, M., Moreno-Armendáriz, M.A., Cruz-Cortés, N., Seck Tuoh Mora, J.C.: Modeling prey-predator dynamics via particle swarm optimization and cellular automata. In: Batyrshin, I., Sidorov, G. (eds.) MICAI 2011, Part II. LNCS, vol. 7095, pp. 189–200. Springer, Heidelberg (2011)
14. Sarikaya, S., Weber, G.-W., Doğrusöz, Y.S.: Combination of conventional regularization methods and genetic algorithm for solving the inverse problem of electrocardiography. In: 2010 5th International Symposium on Health Informatics and Bioinformatics (HIBIT), pp. 13–20 (2010)
15. Cary, S.E., Throne, R.D.: Neural network approach to the inverse problem of electrocardiography. Comput. Cardiol. **1995**, 87–90 (1995)
16. Jiang, M., Xia, L., Shou, G.: The use of genetic algorithms for solving the inverse problem of electrocardiography. In: 28th Annual International Conference of the IEEE Engineering in Medicine and Biology Society, EMBS '06, pp. 3907–3910 (2006)
17. Angeline, P.J.: Evolutionary optimization versus particle swarm optimization: philosophy and performance differences. In: William Porto, V., Saravanan, N., Waagen, D.E., Eiben, A.E. (eds.) Proceedings of the 7th International Conference on Evolutionary Programming VII (EP '98), pp. 601–610. Springer, London (1998)
18. Krink, T., Vesterstrom, J.S., Riget, J.: Particle swarm optimisation with spatial particle extension. In: Proceedings of the 2002 Congress Evolutionary Computation (CEC '02), vol. 2, pp. 1474–1479. IEEE Computer Society (2002)
19. Cui, Z., Chu, Y., Cai, X.: Nearest neighbor interaction PSO based on small-world model. In: Corchado, E., Yin, H. (eds.) IDEAL 2009. LNCS, vol. 5788, pp. 633–640. Springer, Heidelberg (2009)

20. Kennedy, J.: Small worlds and mega-minds: effects of neighborhood topology on particle swarm performance. In: Proceedings of the 1999 Congress on Evolutionary Computation, CEC 99, vol. 3 (1999)
21. Cervantes, A., Galvn, I.M., Isasi, P.: AMPSO: a new particle swarm method for nearest neighborhood classification. Trans. Sys. Man Cyber. Part B **39**, 5 (2009)
22. Akat, S.B., Gazi, V.: Particle swarm optimization with dynamic neighborhood topology: Three neighborhood strategies and preliminary results. In: Swarm Intelligence Symposium, SIS 2008, pp. 1–8. IEEE (2008)
23. Aster, R.C., Borchers, B., Thurber, C.H.: Chapter Four - Tikhonov Regularization. In: Aster, R.C., Borchers, B., Thurber, C.H. (eds.) Parameter Estimation and Inverse Problems, 2nd edn, pp. 93–127. Academic Press, Boston (2013)
24. Sundnes, J., Lines, G.T., Cai, X., Nielsen, B.F., Mardal, K.A., Tveito, A.: Computing the Electrical Activity in the Heart. Monographs in Computational Science and Engineering, vol. 1. Springer, Heidelberg (2006)
25. Wang, D., Kirby, R.M., Johnson, C.R.: Resolution strategies for the finite-element-based solution of the ECG inverse problem. IEEE Trans Biomed. Eng. **57**(2), 220–237 (2010)
26. Wang, Y., Rudy, Y.: Applications of the method of fundamental solutions to potential-based inverse electrocardiography. Ann. Biomed. Eng. **34**(8), 1272–1288 (2006)
27. Wang, D., Kirby, R.M., Johnson, C.R.: Finite element discretization strategies for the inverse electrocardiographic (ECG) problem. In: Dössel, O., Schlegel, W.C. (eds.) WC 2009. IFMBE Proceedings, vol. 25, pp. 729–732. Springer, Heidelberg (2009)
28. Wang, Y., Yagola, A., Yang, C.: Optimization and Regularization for Computational Inverse Problems and Applications. Higher Education Press, Beijing (2011)
29. Charulatha, R., Rudy, Y.: Electrocardiographic imaging: I. Effect of torso inhomgeneities on body surface electrocardiographic potentials. J. Cardiovasc. Electrophysiol. **12**, 229–240 (2001)
30. Gavgani, A. M., and Dogrusoz, Y. S.: Use of genetic algorithm for selection of regularization parameters in multiple constraint inverse ECG problem. In: 2011 Annual International Conference of the IEEE Engineering in Medicine and Biology Society, EMBC, pp. 985–988. IEEE (2011)

Author Index